RECORDING AUDIO

ENGINEERING IN THE STUDIO

BARRY R. HILL

Rivershore Creative

Published in 2019 by Rivershore Press. Visit us at www.rivershorecreative.com.

ISBN: 978-1-7321210-2-7

The audio examples are available at www.rivershorecreative.com

❀ Created with Vellum

CONTENTS

INTRODUCTION

Another book on audio? Really? For heaven's sake, why? There are shelves full of books that cover various aspects of engineering. For years I sifted through the pile, looking for something that seemed appropriate for my introductory recording classes. Some of these are rich resources, full of theory that every engineer should have at hand. But none of them really worked for helping a newcomer get a big picture of what's going on, carefully leading them through an ever-deeper exploration of the concepts and applications.

This book is the result, carefully designed to help you learn how to record and produce audio, use the various tools and techniques we have available to us, and understand some of the theory and concepts behind everything. We'll save the theory for later, though, because it won't mean much until you've spent some time actually recording and using the gear. You should try to get your own setup to work with, even if it's just free software and a single microphone. Experiment and follow along as much as you can so the reading makes more sense. Hands-on is always best, so go back and forth and it'll gradually become clear.

We'll start by walking through a one-track recording, showing the equipment, procedures, and how to make it sound good. If you've already done some recording you could skip over to the music production section, but otherwise this will give you a good step-by-step of what's involved.

Next we'll jump into producing a song, beginning with the mix process. This may seem strange since it's not the first stage of a project, but it's much easier than trying to get a tracking session up and running with the musicians standing around waiting impatiently. Mixing involves use of all the equipment and software in the control room and facilitates learning signal flow. It also provides an end-appreciation for why you should record things right to begin with. After you've gotten the gist of mixdown and how the gear basically works, we'll step through a tracking session, then discuss editing and mastering for when your mix is done. Along the way I provide lots of helpful solutions and suggestions for things you're going to run into, so even if you're familiar with how to track a guitar you'll probably find something useful.

Although making records in the studio is the first thing that comes to mind, there are lots of different applications (and careers) for audio engineers. We'll touch on just a few, such as how to produce a podcast or record audio for a TV show. Music educators are interested in recording their ensembles as well as helping students produce their own songs, so I'll lay out some of the ideas I share with teachers in our graduate program at the college.

Once you understand how the process works and have a functional familiarity with the equipment, we'll go back and explain the specifics in greater detail later in the book—recording console functions and signal flow, microphone design and technique, signal processing, and other audio fundamentals.

There are no shortcuts

Many people these days think all they have to do is get signal into Pro Tools and work the plug-ins to be an engineer. Hardly. If you want to be a real engineer, one who really understands this stuff and can work like a pro for years to come, then you need to think deeper. The goal is not learning how to use a specific recording console or software package. If you learn the standard features, controls, and operations that all recording systems have in common, then you can learn how to use any console or DAW (digital audio workstation). Simply look for those standard patterns and see where they are on that particular system. To do this you need to spend adequate time reading this book, listening to the audio examples, and practicing on your equipment. If you don't, you are taking shortcuts that will prevent you, or at the very least delay you, from learning the essence of audio production. Not understanding these concepts will limit your abilities and creative potential, and will hinder you from adapting to different or unexpected circumstances. I see it every year with students no matter how much we warn them, so dig in and get to work.

You don't have a recording console? No outboard gear? Is this applicable to your computer-based recording system? Absolutely—the basics always apply. If you learn the generic concepts, you can transfer that knowledge to any system, whether it be a large format analog console or a laptop-based Pro Tools rig. The equipment and end product may change over time, but the fundamentals and artistic concepts never do.

You gotta have ears

You can't learn audio without actually listening. The website has tons of audio examples that are described throughout the book. Make sure you use high quality speakers or headphones to listen with; many of these are tough to hear unless you have decent gear—most earbuds won't cut it. Also keep in mind that it takes practice and experience to train your ears, so at first you may not notice some of the settings and nuances. Keep at it!

The fundamentals never change whether you're tracking an album, producing a podcast, or mixing the soundtrack for the latest blockbuster film. If you develop skills at miking, signal flow, using processors, mixing, and most of all—hearing, you can apply that to anything you run into. Good audio is good audio, no matter the situation, so plunge in and enjoy the ride.

ONE
FROM MICROPHONE TO MP3

Let's get you started and walk through a basic recording. Along the way we'll discuss the equipment you need, procedures to follow, and how to make it sound good. You'll then have a pretty solid understanding of what's involved so you can move on to more complex projects.

Recording a track

Recording a source involves a microphone, an audio interface or microphone preamplifier, and a recording device. The goal is to select a mic that works well for that particular source, put it in a spot that sounds best, and get a solid recording signal into the recorder.

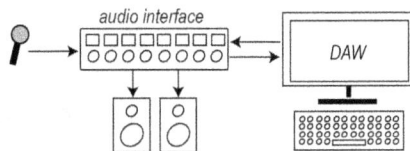

Equipment needed

- Microphone
- Mic stand
- Mic cable
- Audio interface or microphone preamplifier
- Recorder
- Headphones/monitors

Microphone

Probably the best general purpose mic to start with is a large-diaphragm condenser. Most microphones are designed either as a *dynamic* or *condenser*, which signifies how they operate internally. What this means to you for now is that condensers are brighter, more articulate, and can work on most anything. Dynamics are great for drums and other things, but maybe not so much for vocals, piano, acoustic guitar, etc. So, have your local music store help you find an inexpensive model; you can get these in the $100—$300 range (and, of course, way beyond that). Large-diaphragm mics have a larger capsule inside the screen, as opposed to what are commonly called "pencil" condensers. The advantage is a broader frequency response, meaning it picks up lows and highs really well.

Check the microphone stand and make sure everything is tight—adjustable boom arms, the clutch where you raise and lower it, and the base. I've found many a stand where the boom was about to fall off with a mic on it, so be safe.

Don't try screwing the microphone onto the stand—it's too easy to drop it. Do it the other way by loosening the last boom arm segment, hold the mic firmly in one hand, and spin the boom into the mic clip. Don't force it—make sure the threads are lined up properly. To adjust the stand height and angle, first loosen the knobs and tighten them when you're done.

Point it in the right direction. Every mic has an on-axis point which should face the sound source you're miking. Sometimes this is obvious,

other times not so much. For example, long thin mics like the Shure
SM57 and 58 are pointed with the end of the grill facing the source.
Large diaphragm mics usually pick up from one side of the grill; look for
the manufacturer's nameplate or use your ears to make sure. When a
sound is picked up away from the on-axis direction of the microphone, it
will have an unnatural or distant sound quality, which generally doesn't
sound very good.

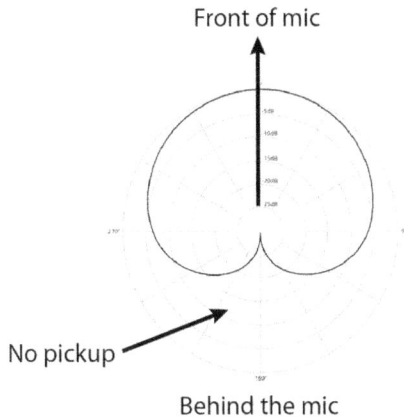

Front of mic

No pickup

Behind the mic

Where to place the mic? This depends on what we're recording, but stay
fairly close, maybe a foot or two away. If it's a vocal (spoken word or
singing), place it at the level of their mouth, perhaps a tad higher, and
angled downward just a bit. Too close in front of the mouth and you'll
get puffs of air which cause problems. If you're getting too many "pops",
place a pop filter between the mic and the mouth. You can also move the
mic over to the side a little so it's not directly in front.

For acoustic guitars point the mic toward the 12th fret (count 'em),
which splits the difference between the neck (finger noise) and the hole
(boominess). I've got lots of miking suggestions in the tracking and mic
technique chapters.

Check the switches on the microphone. If your mic has switches, it will
be one or more of the following:

- *Polar pattern select.* You generally want the mic to pick up sounds from the front only, not from the back or sides. Set the polar pattern switch to the heart-shaped symbol; this represents a cardioid pattern and means the mic will pick up sounds mostly from the front.

- *Attenuation pad.* This reduces incoming signal level in case your sound source is really loud and is overloading (distorting) the mic. Ever put your ear one inch away from a snare drum? For now, leave it off (0dB).
- *Low-cut filter.* This attenuates (reduces) low frequency sounds such as rumble, vocal pops, and trucks driving by. All consoles and DAWs have low-cut filters, so we'd rather set these later while mixing. The symbol looks like a division sign; set the switch to the flat position.

Grab a mic cable and connect it to the mic, making sure you hear a click that indicates it's locked in. Microphone cables have different connectors at each end. The male end with three pins gets plugged into the audio interface or microphone preamplifier (audio follows the direction of the pins). The other end has three holes and goes into the microphone. See the little switch on the connector? Make sure you push this when unplugging the cable. You don't need to wrap the cable around the stand like a mummy. Just do a couple wraps around the stand so the cable doesn't hang out and trip somebody. Once it reaches the floor, run the cable in a way that minimizes getting stepped on by people—this damages the tiny wires inside the cable, which is a bad thing.

Be careful:

- Don't leave a mic on the floor. You just might be the loser who steps on it.
- Don't connect, disconnect, or move microphones when the recording channel is on. It pops through the system, possibly damaging your speakers or headphones.

Room acoustics play a major role in what you're getting in the mic. A plushly-furnished living room sounds much more subdued than a rever-berant bathroom. Spoken word requires a tightly-controlled acoustic environment, so find a smaller room with lots of drapes, couches, book-shelves, or acoustic panels on the walls. Make sure there's not a flat, unfurnished wall close to the person or microphone; this causes direct reflections back into the mic and sounds bad.

Music requires some room ambiance to give it life, so we're generally seeking a balance. Uncontrolled reflections and too much reverberation cause issues such as muddiness and phasing, but a dead-sounding room is terrible. Recording studios should be designed with this in mind, treating walls and ceiling surfaces to control reflections without overly deadening the space. If you're recording in your bedroom, office, or band room at school, experiment in different locations in the room, adjusting materials and furniture, and so on. More ideas and issues with acoustics are presented in the acoustics chapter.

Audio interface / preamplifier

Some microphones are designed with a USB output, rather than a stan-dard cable connector. This is really handy as you can plug it directly into the computer. These don't sound quite as good, though, but for average use they'll work fine. All other recording mics have the 3-pin connector we described earlier, called an *XLR*. The signal output of these microphones is very low and requires a special amplifier to increase it so it's compatible with a console or recorder. Microphone preamplifiers are found on each channel of a recording console or live mixer, some external flash recorders, and audio interfaces.

An audio interface is the bridge between microphones and computers. Plug the mic cable into the XLR input on the interface, then USB (or whatever) to the computer. This device handles the analog to digital conversion for recording, then reverses this when playing back through headphones or speakers connected to the interface's monitor outputs. This is a fairly complex, crucial operation, and as such you generally get what you pay for. Cheap interfaces will sound, well, cheap, so as the budget allows consider investing in a quality interface.

Some audio interface units feature a single microphone input, some two, and so on. If you're planning to record multiple tracks at the same time, such as for a band, you'll need an interface with several mic inputs.

Typical settings on the interface to look for:

- *Headphone/monitor volume*
- *Input level*: sets incoming microphone level for recording. Leave this all the way down for now.
- *+48V*: Phantom power on/off switch. This is required for condenser mics, so turn it on.
- *Pad*: This switch will attenuate incoming signal, such as when the source itself is too loud for the input even when turned down all the way. Leave off for now.
- *Low-cut filter*: Leave off and use the ones in your DAW or console.
- *Source/Mix balance control*: Set to mix so we're hearing only the output from the recorder.

Recorder

This could be a software DAW (digital audio workstation such as Pro Tools or Logic Pro), portable flash recorder, or a washing-machine size multitrack tape recorder from the 80s (awesome, by the way). Let's assume a DAW, so open a new session and set as follows:

- Bit rate: 24 bit
- Sample rate: 44.1kHz
- BWAV or WAV

In the audio settings menu find how to select the audio interface hardware you're using. This will tell the DAW what inputs and outputs are available on the device as well as how to play back your tracks through monitors or headphones.

If it's an empty session, create a new audio track. Set the track's input to the audio interface channel where your mic is connected. For example, a stereo interface will have two mic inputs, so assign track #1 in the DAW to mic input #1.

Arm the track, meaning set it to prepare for recording; it should turn red somewhere. Have the person at the mic begin speaking or playing and gradually turn up the input level on the interface. The DAW track fader won't have anything to do with recording level, so just leave it alone. Watch the meter on the track and set the input level so you have a healthy signal that peaks in the upper region, but well away from the very top. Hitting zero in digital, which is as far as it will go, will turn into instant hash as the system runs out of bits to encode the signal.

That's it—press record and do a take. Keep an eye on the meter; musicians tend to get more excited during an actual take than when getting a level check. Don't do any abrupt level changes, but if it's peaking out you'll have to do another take with a lower level (turn down the interface mic gain control).

If the artist wants another go at it, return the DAW transport to the beginning of the session and just hit record again. It'll save each take you record; look for the clip list window to see everything that's created in the session. One huge tip is to name each audio track as soon as you create it—not later during mixing. As you compile multiple takes and do editing on a track the clip list will automatically use that track name, which is hugely helpful in looking for the needle in the haystack of a large, complex session.

Processing the file

For now we're going to assume a simple one-track recording that we now need to edit and mix into a final audio file. Here are the steps:

- Editing
- Signal processing
- Bouncing/exporting a final file

Editing

In this case, editing should be as simple as getting rid of extra space at the beginning and end of the take. For album production, a single vocal track might be comprised of several takes, where the engineer selects portions of each take to build a final track. An entire chorus can be duplicated to replace another chorus that didn't go so well. We'll leave that for now and stick to our awesome one-track masterpiece.

Editing in a DAW generally doesn't change the actual audio that was recorded; every edit, fade, or track positioning is stored by the software, *describing* what should happen when the final file is output. This provides nearly unlimited flexibility for experimenting and getting it just right.

Look at the beginning of the waveform and you'll see blank space before the performance starts. Select the trimmer tool, hover near the left edge of the waveform, then click and drag over to the right until it's close to the edge of where the audio starts. Now get the selector tool and high-light the remaining narrow region before the waveform; apply a short fade-in so you get a smooth start for the audio. Zoom in as necessary to make it easier to see and select audio.

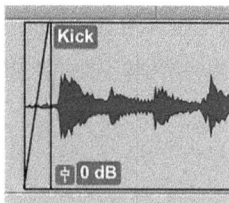

Pro Tools users should get familiar with *smart tools*, which means as you hover the cursor around the waveform different tools will appear. This is an incredibly efficient way to work: hover near the edge and the trimmer tool is selected, top half of the waveform gets the selector tool, bottom half is the grabber tool, and near each top corner pops up a fade tool. To enable smart tools, click the bracket over the set of main tools in the top window.

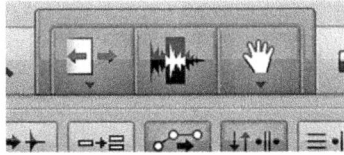

Now trim the end of the audio and apply a smooth fade out. Make sure you're not grabbing too much of the audio in the fade, unless it's an intentional fade out for a song.

Signal processing

Recorded audio nearly always needs some type of processing to correct problems and/or make it sound better. We'll dig into this much more in the music production section, but let's try a few basic things for the track you just recorded.

Filter/EQ

I nearly always put a low-cut (high-pass) filter on a majority of tracks in a session. Filters attenuate (reduce) energy in a particular region. Most instruments and voices do not extend all the way down into the lower frequency ranges, which is where we find miscellaneous noise, rumble, drum leakage, and so on. Click on a track insert and select an EQ, such as the 7-band EQ that comes with Pro Tools. Turn on a low-cut filter and adjust the frequency up the scale until it begins to cut out the lower portion of the sound. Backtrack a tad and you're set. The other control, *slope*, determines how abruptly it will chop off this frequency range. For now, set it for a moderately steep slope (12dB/oct).

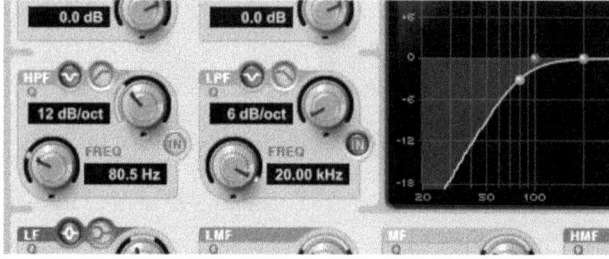

Audio example 1: Low cut filter

The next step is to clean up the mud or cloudiness that's common on acoustically recorded tracks. Find the low-mid band and turn it up 6dB or so. Sweep the frequency select control back and forth between 250 and 400Hz and listen for a spot that sounds more muddy or cloudy than the rest. It'll all sound weird, and it'll take lots of practice and listening to get a feel for it. Once you zero in on this, turn the level control down to -3, -6, or even more as necessary. Bypass the EQ and listen to the before and after; it should be clearer with the EQ turned on. This same approach can be used for an annoying ring, edgy guitar or trumpet, and so on.

There's a third control on most EQs called *bandwidth*, or Q. This controls how wide of a region the EQ will change. Keep this fairly narrow for this step so it doesn't pull out too much of the overall sound of the track.

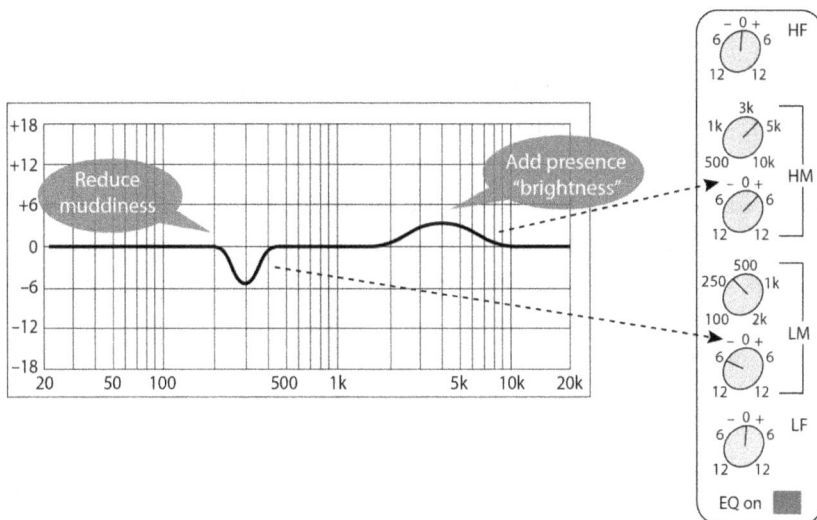

Now that we have a clean, basic sound, let's shape it a bit so it has more presence and sparkle. Grab the hi-mid control and boost it 3dB. Set the Q for a wide bandwidth and sweep around the 4k region. This is a good area for presence and intelligibility; don't overdo it, so perhaps a couple dB is all that's needed, depending on the source. Now try a very gentle boost with the high frequency control. Set it to shelving mode and sweep up around 8-10kHz. This applies a flat boost to the entire upper frequency range, providing more "air" and openness. Lastly, try boosting the low-frequency control a couple dB around 100-250Hz. Exact frequency regions will depend on the source, so over time you'll get a better feel for how to treat different instruments or voices.

Audio example 2: Finding EQ settings

All EQs generally do the same thing, but in different, often subtle, ways. Circuit design, components, and internal settings distinguish one from another, so experiment and try various EQ plugins on your track. Maybe it'll all sound the same now, but over time you'll get a sense of how each makes a unique impact.

Compression

Contrary to popular opinion, there's no law that requires putting a compressor on every track. But, we generally like to compress audio because it can sound tighter, punchier, and fuller. Done wrong, though, it'll dull the tone and crush the life out of a musical part. Along with improving the overall sound, compressors are useful for managing swings in dynamic range (soft to loud). A vocalist singing normally through the verse can go inexplicably insane on the chorus, making it difficult to fit into a mix.

Let's accomplish both of these goals for our one-track project. Insert a compressor on the track and set the ratio to 3:1, fairly fast attack, and a medium release. Now play the track and adjust the threshold down until you see a few dB of gain reduction on the meter. Lowering the threshold will compress more of the track's dynamic range whereas a high setting will only affect the peaks (louder portions). Once the threshold tells the compressor to start, the ratio determines how much it will reduce the output. So a 3:1 setting means that only 1dB will be output for every 3dB coming in. Want to crush a metal singer? Crank the ratio up to 9:1 and lower the threshold until he turns into indecipherable gibberish.

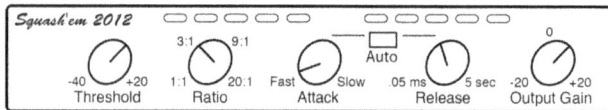

Audio example 3: Compression (light, then heavy)

As with EQs, compressors come in many flavors. Try each one in your DAW or rack and start getting a feel for how they react and sound differently.

Reverb

Now let's give the track some ambiance, such as a musical club or small performance hall. But, unlike most novices who slap effects plugins on all their individual tracks, we're going to set this up properly based on long-standing studio practice.

Click a send on the audio track and select any of the internal buses (1-2 is fine). Turn up the send fader to zero. Now create a new stereo aux track, which is different from an audio track. Set the source for this track to the same internal bus you routed the send to (bus 1-2). This brings a copy of the original audio over to the aux track; if you had several tracks, you could send all of them to the same aux track, thereby adding the same reverb sound to everything.

Insert a reverb plugin on the aux track and pick a preset such as small hall, plate, etc. Play the track and you'll hear reverb; adjust the aux track fader level to balance the amount of reverb with the original track. It's good practice to rename the buses as well, so instead of "bus 1-2" it would be "vocal reverb". In Pro Tools this is done under the IO Setup menu.

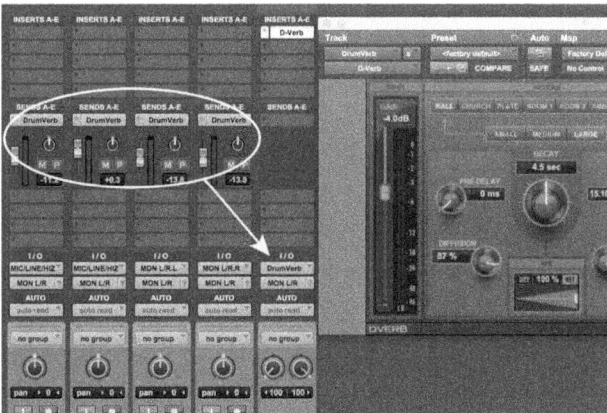

Audio example 4: Reverb (dry, wet)

Final steps

Once you've got things sounding like you want, let's get ready to export it to a new audio file. Create a stereo *master fader* track; drag it over to the right side if it shows up somewhere in the middle of your tracks. All individual tracks now feed this master bus, and you'll see the overall signal level on its meter. This is crucial so that we can be sure the final mixed signal isn't running too high and risking distortion. Play the track at the loudest point and see how high the meter goes; pull down the master fader a bit if necessary.

One last thing to try is inserting a bus compressor or peak limiter on the master fader. Bus compressors are set for gentle compression so as to pull things together a bit. Peak limiters are overused these days, but the idea is to take lower-level audio and increase it while maintaining a ceiling on the overall track. By squashing dynamic range into a smaller space the overall mix sounds much louder and perhaps fuller; overdo it and it sounds distorted and unmusical. For a spoken word podcast a peak limiter can increase intelligibility for listening in the car by keeping the overall volume more consistent.

Pro Tools comes with a peak limiter called Maxim; set the output level for -1dB and slowly decrease the threshold while playing the track. Listen as it gets louder, then progressively more nasty. Find a balance that works and you're done.

Remember, everything you've done in the DAW so far only describes how the recorded audio is to be played back. All of the processing now needs to be finalized in a separate audio file. This is called *bouncing*, so go to the file export menu and set things as follows:

- Bounce source: Main/Mon LR
- File type: WAV
- Format: Interleaved/stereo
- Bit depth: 24
- Sample rate: 44.1

The offline option is non-realtime, meaning it will render the file much quicker (a good thing). If you want an mp3 for sharing, check the box and it'll create both file types. WAV files are higher quality as they are uncompressed, just like a CD. Mp3 files are lower in quality, but feature small file sizes. Set the mp3 option for 256kbit/s.

That's it—you've produced a complete recording from microphone to mp3. All of these steps will come up again in the following chapters, but with more detailed information, options, and recommendations.

TWO

MIXDOWN: SETUP

The goal for mixing is to blend individual tracks from the multitrack recorder so they sound like they all played in the same room, at the same time. To do this we use faders (level balance), pan pots (left-right imaging), and signal processors (tone, levels, effects). Once you get it right, it's saved as a stereo file or, if you're lucky enough to have one, recorded on a 2-track tape recorder. First, let's look at the overall signal flow involved in mixing, both for DAW-only systems (mixing on your laptop) and console-based studios.

Signal flow for mixdown

You'll see the term *signal flow* a great deal in this book. Signal flow is fundamental to understanding how the gear and equipment operate and work together, even on a DAW. Where exactly is the signal flowing from your mic input to the recorder? Will changing an EQ setting affect the compressor? Will the musicians hear an EQ change in their headphones? Did you just lose signal somewhere? Understanding where everything goes from one point to the next is crucial for solving problems as well as coming up with creative ideas. We'll cover lots of each throughout the book, but for now let's give you a mental picture of how things flow for a mix.

Console mixing signal flow

Recording studios have either a multitrack recorder or DAW that's connected to the console. We'll treat both of these as the same thing for now—it's simply the recorder playing back your tracks. The outputs of the multitrack are usually permanently connected by cables to the console input channels in numeric order (multitrack track #1 = console channel input #1). The individual signals flow down each console channel and are all routed to the mix bus fader, which is a combining amplifier that mixes all the individual tracks together. This is the final result that you'll save as a mix file or record to an external 2-track recorder. Along the way you can apply signal processing such as EQ, compression, and reverb, either to individual channels or to the entire mix.

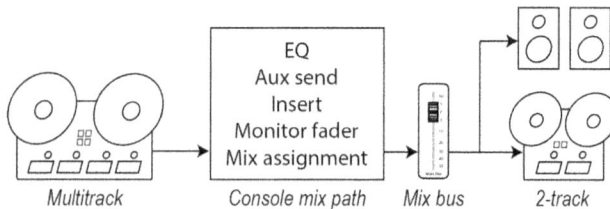

| EQ |
| Aux send |
| Insert |
| Monitor fader |
| Mix assignment |

Multitrack   *Console mix path*   *Mix bus*   *2-track*

If you're playing back from a DAW through the console, the computer has to first connect to a multichannel audio interface. This device converts the audio information to analog signals which are then fed individually to the console channel inputs.

DAW mixing signal flow

If you're mixing on a DAW with no console, signal flow follows the same concepts. The main difference is that it's all virtual—you don't have to plug in any cables. Signals on individual tracks all go to the master mix bus (*master fader* in Pro Tools). You can add processing to a track or the entire mix. Your setup might include a control surface, which has physical faders and controls, but we'll treat this just like a virtual DAW studio as it's not the same as an actual recording console.

We'll step through all of these operations in just a bit, including some creative ways of routing signals. For now, let's jump in and start mixing. Grab some tracks somewhere, import them into your DAW, and get the coffee brewing.

Mixing on a console

Console channels

Turn up the fader for level

At the bottom of each channel, turn up the fader to around unity (indicated by U or zero). Unity is optimum signal gain through a circuit, so it's a good starting point. This gives you room to move the fader up or down as you balance all your tracks.

Route to the mix bus

Now, either at the bottom or near the top of the channel you'll see a group of numbered switches. This is the *assignment matrix*, used for routing incoming signals to some destination such as the mix bus or a track on your recorder. We need to send each channel's signal to the mix bus, so push the button labeled *mix*, *L-R*, or something like it.

Label the channels

Run a strip of label tape across the bottom of the channels and write down the name of each track. This way it's not just "channel 5", but rather "kazoo". Much more descriptive and helpful when you're busy building the mix.

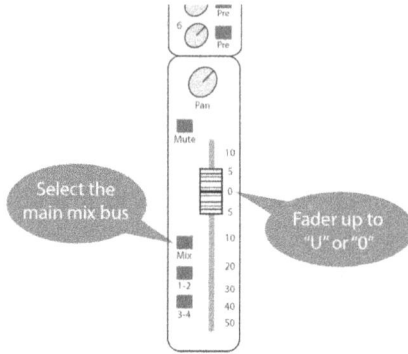

Master mix bus fader

On the right side of your console (or middle, depending on how large your board is), you'll see the master section. In addition to the main mix bus fader, you'll see controls for setting volume, selecting what you want to listen to in the room (mix bus, external 2-track, CD), selecting between different sets of speakers, and other functions. Turn the mix bus fader up to *unity*, which should be labeled zero and is usually all the way up. This is the final level control for everything in your mix and feeds your 2-track recorder. Keep an eye on the meters; you want them up near the top, but definitely not flashing red, which means you're overloading the bus amplifier and distorting your signal.

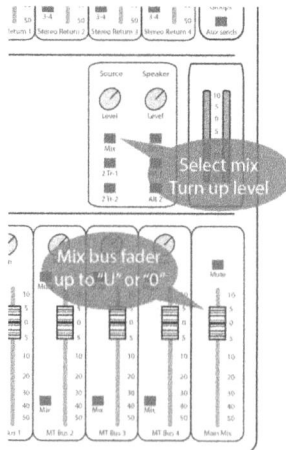

Monitoring source & volume

At this point you have signals coming into each input channel, all routed to the main mix bus and out to the 2-track recorder. But, we still have to route this to our monitor speakers. Recording consoles provide the capability for monitoring different sources in addition to what's going to the mix bus. External media devices such as CD players, iPhones, or 2-track tape recorders can be connected to this master monitor section. Their signals don't go through any channels; this just allows you to hear what's on them.

Find the *source select* buttons and choose the main mix. It might be labeled *mix, main,* or *L-R.* Press play on your multitrack, slowly turn up the control room volume, and you should have sound. This is strictly your listening volume and has nothing to do with setting or changing recording levels. So when the phone rings you can turn this down without messing up the mix that's running to your 2-track.

Be conservative with your volume levels. You need to have a healthy volume to mix, but really loud listening over time is not good for your ears. Find a sound pressure level app (SPL) for your phone and dial it in around 85dB. Keeping the volume down, at least at first, is also a good idea in case you do something wrong that might blow your speakers. Figure out what you're doing first just to be safe.

The multitrack recorder

During mixdown, the multitrack recorder simply plays back the separate tracks recorded during tracking sessions. You need to know where your songs are and be able to find specific locations within each song. For analog recorders, a *track sheet* for each song should have been completed during tracking; this indicates where the song is on the reel, what's on each track, and other helpful information about the recording. Tracks in DAW sessions should be named and include any relevant notes.

Hear Real Good Studios
101 Analog Way
Deaf Valley, NC 99999

Client:	
Producer:	
Engineer:	
Date:	Speed:
Studio:	Level:
NR:	Tones:

Title:				Time:		Cut #		Reel:
1	2	3	4	5	6	7	8
9	10	11	12	13	14	15	16
17	18	19	20	21	22	23	24

Cues:	Notes:

Prepare the analog multitrack recorder

Clean the tape path on the machine—recording tape sheds its particles, which negatively affects sound quality and machine performance. Analog machines need cleaning everywhere the tape touches. This includes the heads, rollers, tension arms, and other tape guides. You should use denatured alcohol or a blend with very low water content for metal parts. Rubber rollers need a general household cleaner that won't dry them out over time. I've provided more instructions in the chapter on analog tape recording.

Load tape and rewind to the beginning. Reel-to-reel tapes should always be stored without rewinding after playback, referred to as *tails out*. Reset the tape counter and fast-forward to the desired song location. Put the machine into reproduce mode (sometimes labeled *repro* or *tape* mode). This is the highest quality playback mode for analog recorders.

Prepare the DAW session

If they're not already defined, set memory locations within each song that will allow you to quickly jump to a section, such as intro, first verse, and final chorus. Re-arrange track order to organize instruments and voice parts into logical groupings, such as guitars, drums, backup vocals, etc. I have a certain track layout I've used for thirty years and it speeds up my workflow not having to search for stuff while I'm working. Color-coding track groupings is another very effective visual organizer, especially for larger sessions.

Create stereo aux tracks for the reverbs you plan to use, such as a vocal verb, drums, and entire mix. Insert an effects plugin on each of these, pick a basic preset that seems close to what you might want, but don't worry about settings right now. On the tracks for each group, set up sends that feed an internal bus, setting that bus as the source for the aux track. In the IO Settings menu, name these buses something meaningful, such as "vocal reverb", "drum reverb", and so on.

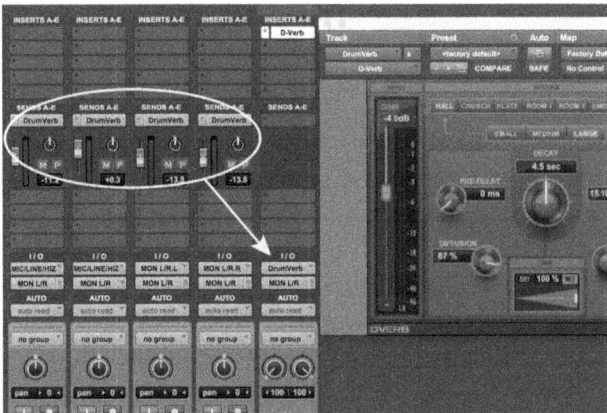

You can also set up aux tracks for group processing, such as compressing the background vocals. For each of your track groups, set up another aux track, bus sends from each track, and insert a compressor. It's the same procedure as you did for reverb, only with a compressor instead. You don't want to use the same send and aux track for both processors; you want total control over each effect individually, including the ability to set fader balance, add an EQ, etc.

Make sure you have a master fader to the far right of your tracks. Everything feeds to the right, so this captures all your tracks and blends them for the final mix. This is crucial for two reasons: the meter indicates what the overall mix levels are doing, and a bus compressor or peak limiter can be inserted on the entire mix. I'd go easy, if at all, on the peak limiter, saving that for a more focused mastering stage. But a bus compressor has been a standard trick for decades; it features gentle compression that pulls, or glues, the mix together, making it sound tighter and punchier. My favorite is the SSL G-bus compressor, set for only a dB or two of gain reduction.

Finally, the more tracks a session has the greater chance it's going to push the overall mix level too high. Instead of starting with all faders set at zero, go to the Groups window, turn on "all groups", and pull a fader down a few dB. All faders will lower; now turn off the group. Your session should now be ready to start mixing.

The two-track recorder

Digital media recorders are fed from the console mix bus, meaning there are analog cables connecting the console mix outputs to the inputs of the recorder. Push the record or monitor button to see incoming signal level. The goal is to get somewhat close to zero (top of the scale) without going over. As long as the device is set for 24-bit recording it's not so crucial to push zero too much. Just get a healthy level.

Analog tape recorders are also fed from the console mix bus. Clean the tape path, load a reel of blank tape, and put the machine into *input* mode. You'll see signal level on its meters and can adjust this either on the recorder or from the console. The meter will be different from digital machines; instead of zero being max, zero is considered unity or optimum level. There are a few dB of headroom above zero, so you can go over within reason.

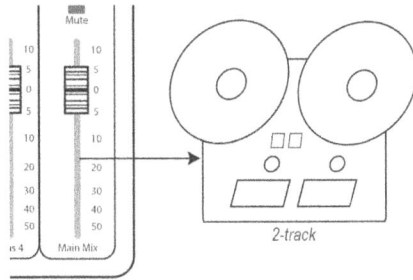

Another option is to mix through a console, but record the final mix in the computer, rather than to an external recorder. The advantage is you now have a digital file in software, ready for final editing and mastering. Analog cables run from the console mix bus to an external 2-channel audio interface, which is connected to your computer via USB. Launch a stereo editor (2-channel recording application) and set its input source to the audio interface. Press record, then go back to the DAW and press play. Switch back to the editor to see signal coming in. Here's the layout:

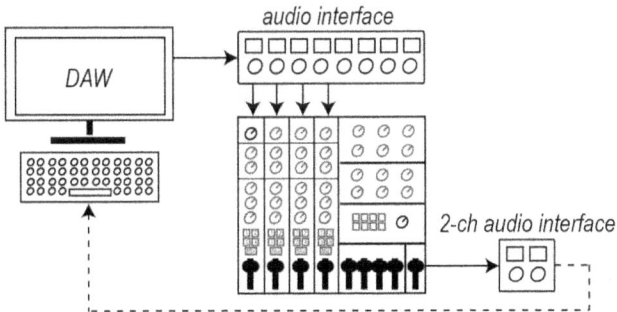

It's possible to use the same interface for both DAW playback and mix recording, but it's easier to understand the setup and signal flow to show it like this.

By the way, you don't have to go insane trying to push play on the DAW and record on the 2-track at exactly the same time. Make sure the 2-track is recording, then start the DAW. At the end of the song make sure it has completely faded away...and then wait a couple more seconds before stopping both transports. The extra silence will be chopped off during editing.

Playing back the mix

Once you've recorded a mix and want to listen back, you have to reselect your monitor source on the console. Remember where you selected the main mix option and turned up the volume to hear your mix from the console? Just push the button that corresponds to your 2-track recorder. Switch tape machines to *repro* mode for playback; a 2-track left in input mode will show no sound or signals, making you think the tape is blank.

A word about levels

When setting levels to tape, there is an optimum range you should be shooting for. If you record too low on analog tape you get lots of tape hiss; too high and you get distortion. If you record too low with digital recorders you get lower sound quality, but if you go over the top you get nasty distortion. Your goal is to average as close to the top as possible without going over. Current digital systems provide enough dynamic range for this to not be so crucial, but you should still try to maintain a healthy level throughout. Experiment with your particular equipment and see how high you can record before distortion begins to kick in.

Mixing in a DAW

Recording a DAW mix without a console is performed internally in software through a *bounce* operation. The stereo mix bus signal is rendered to a new file on your drive, usually as a WAV, BWAV, or AIFF format. MP3 files can also be generated at this time, but always make sure you get a full-quality, non-compressed file such as a WAV. To play the file back, open it in a stereo editor, iTunes, or any other media player on your computer.

That's all there is. No equipment, no cables, easy. You still have to pay attention to the mix bus levels so your final bounced file isn't either too low or distorted. Always use a master mix fader (master fader) and watch levels throughout the song. If it peaks somewhere, lower the master fader a bit or bring down the faders on individual tracks. There are many other variables involved with this that we'll cover later in this chapter and in the rest of the book, but for now just use the fader and keep it from peaking.

Getting a rough mix

Where do you start? Most young engineers immediately start putting plugins on everything and setting automation. Forget that for now. Play the file and just listen. What's there? Even if you're working with just a voice track, such as for a podcast, get a feel for what it sounds like and what it might need.

How should you go about building a mix? Some engineers start with the drums and rhythm section, then add everything else on top. Some start with the lead vocal and build the band around it. It's entirely up to you, but it might be advisable to begin with single instruments and sections until you become proficient at hearing things amidst the jumble of the entire band. It's much easier to listen for EQ settings one instrument at a time—just remember that it will sound different when you blend it with all the other parts. I did a project once where each member of the band spent an inordinate amount of time playing with EQ on their solo'd part, and of course it all went out the window when everything was brought back together. At least I wasn't paying for it.

For now, experiment with different fader levels for volume balance between tracks, then begin placing parts into left-right perspective by turning pan controls. Don't worry so much about EQ at first—just get a decent balance and stereo imaging. Also don't worry about any extraneous noises right before the music starts, such as a count-off leading into the song. These will be edited out after the mix is done.

Audio example 5: Editing count-offs

Think musically about what the song is doing. This determines how various tracks are balanced and arranged in relation to one another. Just how loud should that kick drum be? How loud is the lead vocal in relation to the band? After listening to music for your entire life, now you're the one who has to decide exactly where to place the acoustic guitar. It's harder than you think, and the one constant is that you'll come back the next day and hear it differently.

Listen to see if the parts fit together musically in terms of rhythm, intonation, harmony, and the overall groove. If a part doesn't quite seem to fit in, either it needs some attention or perhaps just needs to be dropped from the song altogether.

Go ahead and print this rough mix and listen back. Your perspective will be different when listening this way because you're not busy thinking about what faders to move during the mix.

Audio example 6: Rough mix with no panning, levels, EQ, or processing

Audio example 7: Mix with panning, levels, and EQ

Audio example 8: Mix with panning, levels, EQ, and effects

Audio examples 9, 10, 11: Getting the guitar right in the mix—third time's the charm

Audio examples 12, 13, 14: Same for the harmonica

Audio examples 15, 16, 17: Same for a vocal

Audio example 18: What's missing?

THREE

MIXDOWN: SOLVING PROBLEMS

Now that you've gotten a basic feel for roughing a mix together, let's make it sound better. Refining a mix involves two main steps: finding and rectifying problems with the recorded tracks and creatively enhancing those tracks. Problems should be dealt with first to make room for creative, productive adjustments. Several common issues you might have to deal with are explained here along with possible ways to fix them.

Problems to listen for

The tracks sound muddy and boomy

All acoustic sound sources have a resonant frequency range, which is increased energy (amplitude) in that band. Tonally it generally results in a muddy sound that will affect clarity and presence of the track. Resonant frequencies are usually in the low-mid range somewhere between 200 and 400Hz, depending on the source. If not removed, the combined effect from all your tracks will result in an overall muddy sounding mix that lacks clarity.

Another cause for a boomy sounding track is when a directional microphone (such as a cardioid) is placed too close to a source during tracking. These mics exhibit an increased bass response in this situation; the solution during tracking is to move the mic back a bit, but during a mix you'll have to EQ it out as best you can.

Example

Muddy or boomy sound in a kick drum, acoustic guitar, or vocal

Solution

Use a parametric EQ to find the offending frequency range

Equalizers come in different design types, such as the bass/treble control on your home stereo or the graphic EQ on the PA system. All EQs will boost and cut frequency energy at specified regions in the spectrum—in other words, you can turn up the treble for brightness or turn down the low-midrange to reduce muddiness. Whereas a graphic EQ is set to fixed frequencies, a *parametric* EQ provides an extra control that allows sweeping around the frequency spectrum to find specific frequencies to boost or cut. Most recording consoles and EQ plugins feature some type of parametric EQ. For each band (high, high-mid, low-mid, low) there will be at least two controls: one to boost or cut the signal level, another to select the desired frequency region. If all this sounds overly complicated, just follow the procedure below as best you can and we'll dig deeper into the details later in the book.

How to find the resonance

Press the solo button on the kick drum channel to single it out. Make sure the overall solo level control, found in the master section of the console, is turned up so you can hear it. Turn on the EQ if your console has an on-switch, or insert an EQ plugin on your DAW. Turn up the gain (+/- knob) for the low-mid EQ at least 6 dB or so; it should now sound muddier. Now rotate the frequency select control adjacent to it and listen for the changing sound of the frequencies as you move up and down the scale. Turn it back and forth until you can distinguish a region which seems to sound more muddy or cloudy. All of it will sound weird, so learning what to listen for will take lots of practice and experience. Now reduce the boost control back to a negative number, at least -3dB, maybe as low as -9. The more you cut, the more of the overall sound will be taken out, possibly resulting in a sound that's too thin.

Some parametric EQs provide one additional control that narrows the region of frequencies being adjusted (bandwidth, or Q). By narrowing the bandwidth of affected frequencies, less of the overall sound will change, allowing you to attenuate (cut) only the offending frequencies.

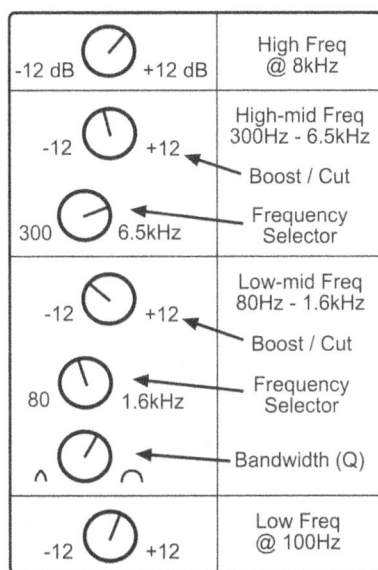

Parametric EQ Module

This same procedure can be used to isolate other EQ problems, such as harsh mid tones from an electric guitar, a 60Hz ground loop, or an overly bright, sibilant voice track. Generally speaking, most engineers will attenuate EQ problems, rather than boosting something to cover it up. For example, if your vocal track is muddy, don't boost the high-mids to brighten it; attenuate the low-mids like we've described here to clean up the problem. This will leave you with a brighter track that has more clarity.

Audio example 19: Acoustic guitar EQ with low-mid cut & upper-mid boost (out/in)

Audio example 20: Bass guitar with low-mids attenuated (out/in)

Audio example 21: Keyboard with low-mids attenuated (in/out) to avoid interfering with other low-frequency tracks in the mix

Audio example 22: Spoken word track cleaned up (out/in)

Audio example 23: Muddy/clean mix

I hear other instruments on one track

If more than one instrument is playing in the studio at the same time during a tracking session, individual microphones might pick up other sounds besides the instruments they are assigned to. This is called leakage. Most of this should be dealt with during the tracking session by further isolating the sound sources from each other, though some amount of leakage can be reduced during mixdown. Leakage is easy to hear; solo the lead vocal track and if you hear drums in the background, you've got a potential problem.

Example

Snare drum sound leaks into the kick drum mic

Solution

Insert a noise gate into the kick drum channel. A noise gate is a signal processing device that can be used to attenuate (lower) or even shut down an audio channel when the signal falls below a certain level, called a threshold. It is commonly used to reduce background noise on a channel when the main audio signal is not present, such as during pauses between phrases, solos, etc. Another application could be an interview, where one person stops talking and the other voice carries over into the mic. Set the gate to lower the first person's track when they're not talking, thereby reducing the amount of the background voice that leaks into the channel.

Getting back to the kick and snare, a lot of drum grooves feature the kick and snare in an alternating pattern: kick on beats 1&3 and the snare on 2&4. For the kick mic, the snare sound will be lower in level, since the mic is much closer to the kick drum. Therefore, the solution is to set the noise gate so that the kick drum will "open the gate"; the quieter snare drum hopefully won't be enough to keep it open, so it effectively shuts the channel down until the next kick.

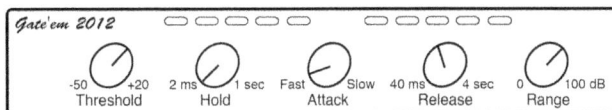

How to insert the noise gate

On the patchbay, take a patch cable and connect the kick drum channel *insert send* to the input of any noise gate channel (a single noise gate unit may have one, two, or more individual channels that can be used on different instruments). With another cable connect the noise gate output to the kick drum channel's *insert return*. For DAWs, simply insert a dynamics plugin on the kick track; this usually includes a compressor and noise gate, individually selectable.

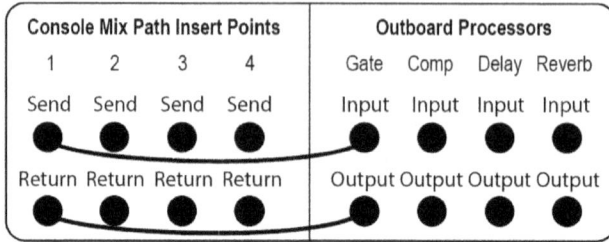

With this patch in place, the kick drum signal is now coming from the multitrack into the console channel input, breaking away and going to the noise gate, then returning to the same console channel to continue the path through the fader and on to the mix bus. In other words it's being detoured out of the console for processing, so what goes to the mix bus is different from the original recording.

So, what's a patchbay? It's a central switchboard for all equipment in the studio; all the inputs and outputs for each device are connected here. To route a signal from the console to an outboard processor, such as a noise gate, simply connect the appropriately labeled jacks on the patchbay. If your studio does not have a patchbay for connecting gear, just use special insert cables and plug directly into your console jacks on the back.

DAW-based studios with hardware processors (compressors, FX, etc) will often have a physical patchbay. This allows you to insert the device using the inputs and outputs (IO) of the audio interface. Once the patching is done, go to the DAW track, open an insert, and select that IO. It's just like inserting a plugin, except that you're now routing signal out of the computer to the device and back. It's the best of both worlds—sometimes you just can't beat a piece of real hardware, but while also enjoying the convenience and flexibility of plugins.

How to set the noise gate

Gates must be tweaked (adjusted) differently for each situation and can be fairly tricky to set. The first two parameters to concentrate on are the threshold and range controls. Threshold determines when the gate will start attenuating the signal (shutting the gate). Any audio signal below the set threshold will be reduced; when the guitar stops playing and fades away it will fall below threshold, whereby the gate attenuates the amp hum and noise. The range, or floor, determines how low the signal will be reduced, so a lower floor will sound like it's shutting down the channel. This needs to be balanced out so you get the desired reduction in leakage without forcing the unit to over-reach during very quick operation.

Next come the attack and release controls. Once a signal rises above threshold, how long will the gate take to open up fully? That's the attack time, and you generally want it as fast as it will go so it opens up immediately. The release time, however, will take some experimentation. If set too fast, the gate will close down quicker. This might cut off your sound before it completely fades away, so in that case lengthen the release a bit. If set too long, the gate won't close down before the leakage becomes evident.

Finally, you might see a hold control. This sets a timer that holds the gate open even if the signal has fallen below threshold. If you've heard a gated-reverb effect (fond memories of the 80s here), that's how it's done. Most of the time, though, just leave it all the way off.

For starters, set the hold to zero, attack time very fast, release somewhere in the middle, and range fairly low. Now play with the threshold and start listening. If the gate is cutting out the kick drum sound itself, then you need to raise the threshold. Experiment to find the appropriate balance; there may not be a perfect fix. Don't forget the attack and release settings as they directly impact how the gate will operate on that particular signal. Wanna have some fun with it? Set both attack and release to super fast and listen to it go crazy.

Audio example 24: Snare before/after gating to eliminate the kick leakage

---

What's all that hum and rumble?

Any continuous background noise that's not part of the music should be reduced as much as possible. This includes guitar amplifier hum, air conditioning rumble, etc. These noises are easily heard and need to be eliminated from the mix. There are various methods for resolving this.

Example #1

Guitar amplifier hum can be heard when guitarist isn't playing

Solution

This is easily solved by inserting a noise gate into the guitar channel in the same manner as for the kick drum described earlier. Simply set the gate threshold just above the level of the amp noise. Set the release time so it doesn't chop off any of the guitar sound, including the fade out of the last note played. When the guitar plays, the gate immediately opens up. When it stops, the gate shuts down, reducing the amount of amp noise.

Audio example 25: Using a gate to eliminate amp noise

Example #2

Air conditioner rumble in the background

Solution

Most rumble caused by air systems, traffic outside the studio, and so on are low frequency sounds. These can usually be reduced by the use of a low-cut filter. As we described earlier, filters sharply attenuate all frequencies below a preset frequency (or above, in the case of low-pass filters). This point, known as the cutoff frequency, is most often set around 75Hz on consoles; plugins let you set it anywhere you want. Turn the frequency up until part of the instrument starts to disappear, then back it down a bit. Find the slope control and set it for 24dB/oct; this determines how abruptly the filter attenuates beyond the cut-off frequency setting.

Of course, filters shouldn't be used on bass guitars, kick drums, and other low-frequency instruments that go down to 30Hz and beyond, but I usually put them on everything else. The result is a cleaner bottom end for the entire mix.

Audio example 26: Low-cut filter eliminates room noise (out/in)

The best solution for background noise is to use plugins designed to remove noise, clicks, hum, and so on. These tools identify the noise and attempt to remove it with little impact on the music or other audio. They've gotten so good at it these days it's almost like magic, and it makes an engineer's life much easier. Don't overdo the settings or it'll introduce weird artifacts to the sound. Experiment to find the right balance between noise and a clean source.

It's difficult to keep a track consistent in the mix

When musicians play, their dynamic range can vary quite a bit between soft and loud passages. Even during a relatively consistent part, such as a rhythm guitar, fluctuations in a musician's playing cause significant differences in recording levels. This makes it more difficult to create an overall balance between all the tracks, because various parts can jump in and out along the way. Non-professional musicians are much more prone to this, whereas a studio musician knows how to keep things consistent and yet retain the musicality of the part.

Example #1

The pianist plays a few big, loud chords during a climax of the song which covers up the vocal

Solution

There are several possible solutions. The easiest might be simply pulling back the piano track faders at this point, as long as it doesn't affect the overall dynamic push of the song too much. If this doesn't work, insert a compressor on the piano track. In simple terms, a compressor is an automatic volume control—like a cruise control for signal level. It reduces the level on audio signals that get too high and cause problems for the overall mix or recording media (causing distortion).

How do you do this?

Just like you did for the noise gate, patch the channel insert send for the offending track to the input of a compressor through the patchbay. Then patch the output of the compressor back into the insert return for the same channel. Remember each channel of a compressor (as well as a noise gate) only affects the single channel it's inserted into. If you are compressing or gating a stereo pair of tracks (two mics on a piano, for example), then you must connect a compressor channel for each track, preferably using a stereo compressor which can link the two channels together.

If you don't have a patchbay for your processors, or in case you're wondering how the patchbay magically makes these connections, the following diagram shows exactly what's happening to get the compressor unit (or any other processor) inserted into a specific channel signal.

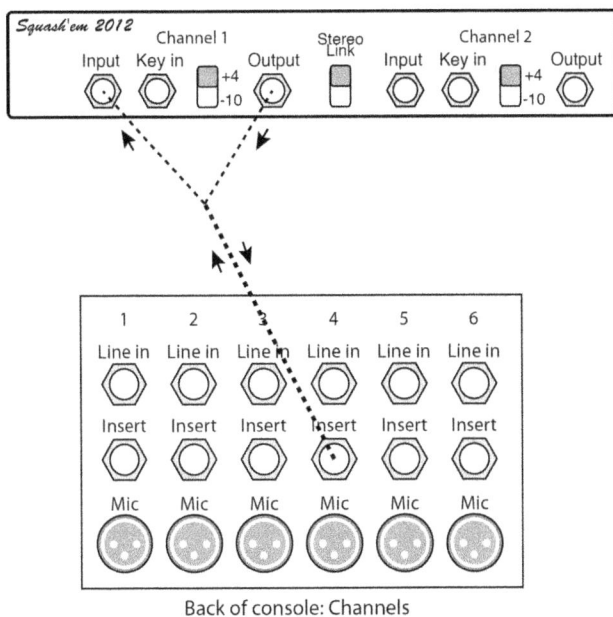

Back of console: Channels

For DAWs, you'll find compressor plugins as well as dynamics, which usually include a compressor and noise gate.

The controls for a compressor are similar to that of a noise gate, but operate in the opposite direction. Signals increasing *above* threshold will be compressed (reduced). The amount of compression is determined by the ratio control; the numeric markings signify the amount of input vs output signal. In other words, a 3:1 ratio dictates that for every 3dB of signal coming in over threshold, only 1dB of that signal is output from the compressor. Therefore, a higher ratio such as 6:1 will provide a greater amount of compression.

If you only need to knock off a few high peaks, but don't want to affect the entire dynamic range, select a higher threshold and higher ratio (6:1 or so). Watch for the gain reduction meter to flash only on loud, strong parts. If the overall signal needs to be compressed to tighten the entire range, set a lower threshold so more of the original signal will be compressed. Lower the ratio if it sounds too squashed; it's a balance between these two controls. If your unit has an auto setting for attack and release, turn it on and let it do its thing. At some point you want to get a feel for these parameters, but for now move on.

Learning compressors and noise gates takes lots of practice and experimentation. Take time to run various tracks through them and get a feel for how they work.

Audio example 27: Guitar compression (out/in)

Audio example 28: Snare compression (out/in)

Audio example 29: Bass compression (out/in)

Example #2

The bass guitar continually jumps in and out of the mix because the recording levels are all over the meter

Solution #1

Insert a compressor into the bass track to smooth out the variation in levels. Try a 4:1 ratio and lower the threshold until the meters indicate 3-6dB gain reduction. Don't overdo it, but adjust until it seems to sit in the mix more evenly.

Solution #2

Hire a professional musician.

Sibilance on the vocal

The consonant "s" is a tricky sound; it inherently tends to stick out more than others. This hissing sound is edgy and annoying, and in the days of tape caused lots of problems with distortion. Often a vocal track gets recorded with excess energy in the 6-7kHz frequency region and needs to be tamed.

Solution #1

Track it properly to begin with. Listen carefully to the vocal mic when getting set for a take; make sure the mic isn't positioned directly downward in front of the mouth. Set it slightly higher and perhaps angled away a bit, even over to the side just a tad. Use a pop-screen.

Solution #2

To fix this during a mix, insert a de-esser plugin. These are essentially frequency-dependent compressors and are fairly simple to use. Set the exact frequency region of sibilance for that particular track, and then tell it how much to compress. Strike a balance so you don't hear the compression too much, but enough so it reduces the annoying sibilance.

What was that I just heard?

During tracking it's inevitable that extra sounds will end up on your recorded tracks. Musicians cough, music stands get hit, cellphones ring, and drumsticks click on things. I once tracked an album that ended up being mixed in Nashville by somebody who didn't pay enough attention to this. One of the vocalists coughed during the instrumental introduction; the engineer muted the vocal tracks during the mix, but somehow didn't notice they were still being fed to the reverb unit.

Solution

Spend time before the mix session reviewing the tracks for noise and other issues. Erase unwanted noises and make notes on other items to remember during the mix. It's advisable to mute (turn off the channel) all unused tracks during the song, such as a vocal part during an instrumental solo. Any channels not used in the song should be kept muted. Not only will muting unused tracks prevent unwanted noises inadvertently finding their way into the final mix, but it will also reduce noise inherent to electronic devices. If you're using analog tape, each open channel playing back from the recorder contributes 3dB of tape hiss (not a good thing, in case you're wondering).

FOUR

MIXDOWN: REFINING

Creative refinements to the mix

Once problems have been identified and remedied as much as possible, you can focus on constructive elements that creatively enhance the mix. This includes equalization, reverberation, special effects, and other tricks.

Tonal adjustments

Tracks might need brightening or a little more bottom end. As opposed to the destructive "fixing" EQ discussed earlier where resonant frequencies and other problems are eliminated, constructive EQ shapes the tonal characteristics of the mix to the engineer's and producer's taste.

Example #1

Cymbals need more shimmer and zing on top

Solution

Boost high frequency EQ on the cymbal tracks. High frequency EQs on a console are usually set around 8–12kHz, which is the upper range of cymbals. Use the same range for a plugin and move around a bit to find what you like.

Example #2

Snare drum needs more snap

Solution

Using a parametric EQ, boost the hi-mid EQ and then sweep the frequency selector around to find the crack of the drum sound. Once the desired region is found, usually around 2–4kHz for a snare, boost the EQ to desired taste.

Be careful about boosting too much. Turning up an EQ band will increase the signal's amplitude, which at some point will overload the circuits in the console or DAW signal chain (distortion). Some people also spend a lot of time EQing each track by itself, only to find that when it's added to the overall mix it doesn't work so well. Other than a few things like this to keep in mind, there are no rules. Just turn the knobs until you like what you hear.

Example #3

Vocal needs more presence

Solution

Presence refers to an open, up-front clarity that helps a track be easily heard in a mix. For most instruments and voice these are high-mids, so boost the high-mid band, sweep around with the frequency select, and find a region that provides clarity in the mix. Adjust the EQ boost where you want it. This works great for acoustic guitar, toms—most anything. Just don't boost everything in the same frequency range or it'll all crash together and sound harsh.

Example #4

It's difficult to hear what the bass guitar is playing

Solution

What I call the definition of a bass guitar lies in the mid-range, where the finger work is happening. Once you have a full bottom end and have reduced any muddiness in the low-mids, boost around 2kHz and sweep around a bit to find note clarity. Again, you might end up boosting more than you'd think so it works in context of the mix.

Audio example 30: Acoustic guitar EQ (out/in)

Audio example 31: Setting the acoustic guitar EQ

Audio example 32: Brightening the cymbals with hi-freq boost (out/in)

Audio example 33: Snare drum EQ adding snap and punch (out/in)

Audio example 34: Drumset EQ on individual tracks (out/in)

Audio example 35: Too much hi-EQ boost on cymbals

Ambiance & reverb

Since tracks are often close-miked in an acoustically controlled studio, there might be minimal natural reverberation that provides music with a sense of space. Music needs reverberation to make it sound natural, so often this ambiance must be added in the mix. The engineer can choose the type of reverberation sound (large concert hall vs small jazz club) as well as how to add reverb to each of the tracks. One overall reverb sound can be blended on all tracks, or different reverb patches (settings on the devices) can be specified for individual tracks. A reverb device simulates how sound waves bounce around in a particular type and size room. Be sure to experiment with the various parameters on the unit—the factory presets are great, but won't work perfectly for each mix situation. These parameters include changing how long the reverb lasts (reverb time), how soon after the initial audio signal the reverb begins to sound (pre-delay), how large or small the reverb sounds (hall vs small club), and others. See the chapter on time-based processors for more detail on how they work and how to use the parameters.

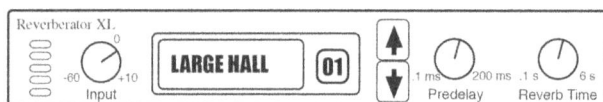

How to do it

Reverb and effects devices are usually not patched into the console the same way as compressors and noise gates, which use the insert sends and returns on individual channels. Reverb devices are instead accessed through the use of sub-mixers in the console called aux sends and returns. Each channel has an identical number of auxiliary sends. Aux send #1 on all channels will feed a common summing amplifier, called the master aux 1 send, located in the master section of the console. The output from this can be patched into an outboard device such as a reverb unit.

The output of the reverb unit must then be routed back into the console and blended into the overall mix. It does not return back to the original tracks. The reverb device output can be connected to an aux return, which is a simple input on the console that directly feeds the mix bus, or it can even be patched into an unused input channel and routed to the mix bus just like the other music tracks. Using an extra input channel gives the engineer control over EQ, panning, and fader level. This can be more powerful than the aux returns on some consoles that may not have these features. The result is the same—the reverb'd signal is then blended with the original tracks at the mix bus, giving the perception of "adding reverb" to the sound. Think of this procedure as a loop where your original signal must go out to the effects unit using an aux send, then must return back to the console using an aux return.

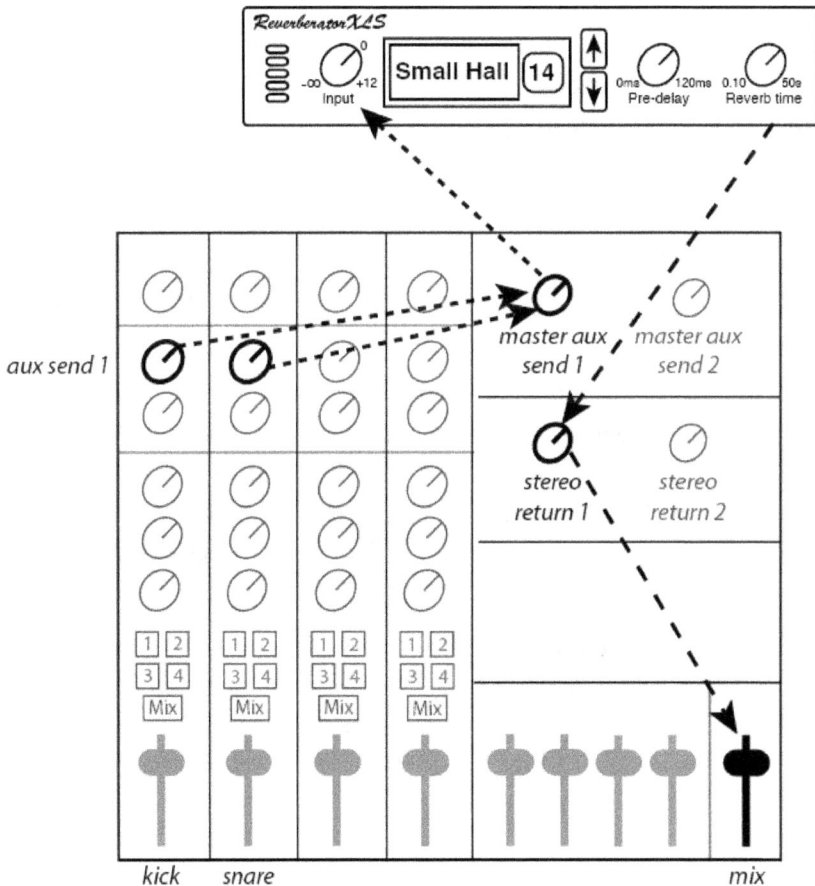

To set all this up, patch the aux send outputs on your patchbay to the inputs of the effects device. Use two aux sends if the device is set to stereo input mode. Now connect the outputs of the device into two aux return jacks (or empty input channels). Note that some consoles have mono returns, others have stereo returns. If your console has stereo returns, then there will be two jacks on the patchbay for this return, but only one level pot on the board that controls both. If the console has mono returns, then two returns must be used for each stereo effects device output. Make sure you pan each return left and right so you preserve the stereo signal from the device.

Now turning up aux 1 (or whichever aux you use) on any of the music tracks will send a copy of those tracks to the reverb unit, which then adds reverb to the signals. If a different reverb patch is desired, say for the vocal track, then simply patch a second reverb unit from a different aux send. Route the output from this device into another aux return or unused input channel on the console.

Always watch your levels. As you send more tracks to the same aux bus, the overall signal level will increase. Make sure you're not overloading the input of the reverb device. If so, turn down the input control on the device, the master aux send on the console, or individual track aux sends.

One mistake people often make with DAWs is inserting the same reverb plugin on each track, using the exact same settings. You might run out of DSP at some point, especially with power-hungry plugins such as reverbs. Now if you're using different settings on various tracks, then go for it—you'll have to insert individual plugins. Otherwise follow the same aux send and return concept we just described.

Use the sends on your tracks to feed a virtual bus. Bring this bus back as the source for an aux track, and insert a single instance of the reverb plugin on that track. Now you can balance the amount of each track that gets reverb and easily adjust the overall reverb level into the mix with a single fader.

Audio examples 36-39: No reverb (dry), large hall reverb, small hall, medium room

Audio example 40: Reverb with distinct echoes

One more item for aux sends. It's usually set by default, but you want to send your aux signals *post-fader* from each channel. This merely determines where in the channel signal path the aux gets its source—either before or after the channel fader. If you send an aux signal before the channel fader level, then any time you move the fader up or down it won't affect the aux level leaving that channel. If the aux is after the fader, then moving the fader does affect the amount of aux signal going out. For adding reverb and most effects you typically want the aux send to follow the fader level, so set your auxes post-fader. In other words, when you use the fader to turn your vocal up and down during the mix, you want the vocal reverb to adjust with it. In the following audio example of a pre-fader send, listen for how the reverb continues after the main channel fader has been attenuated all the way—a nice effect when you want it, but generally not for normal reverb situations.

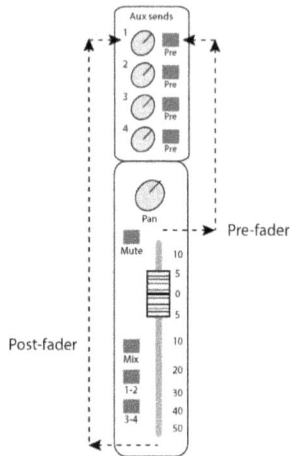

Audio examples 41 & 42: Setting the aux send post-fader vs pre-fader

---

Delay

In the early days of tape it was discovered that introducing a copy of a signal back into the mix at a slightly later time was a very interesting effect, and so the popular application of a vocal slapback delay became a de facto standard over the years. By varying the tape speed of the signal copy, an engineer could change the timing of the delay, and soon devices were developed that did this for you. Now of course we have all sorts of digital processing to generate single delays, multiple delays, and so on. The idea is that the device takes an incoming signal, waits for a specified period of time, then outputs the signal.

Example #1

Add more depth and presence to a vocal

Solution

Add a slap delay to your lead vocal track. This works just like the reverb setup, so you turn up an aux send on the vocal channel, turn up the master aux send, patch this into the delay unit, then return that back into an empty channel or aux return. Set the delay time to around 60ms or so, and try keeping the level of the delay fairly low so it's not so obvious. The idea is to add a presence to the mix, but not be blatant about it. Sometimes, though, you want it to be out there, so see what works for each particular situation.

You can also try adding multiple delays around the vocal, panning them slightly off to each side. Again, keep them low so you can't really hear them, but will miss them when muted. Set delay times to match the tempo and feel of the song or it'll get fairly messy.

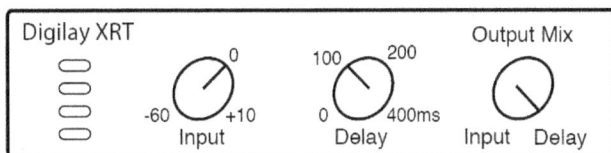

Example #2

Add depth and motion to a track

Solution

Send a copy of an acoustic guitar or organ track to a delay unit with a 15–30ms delay time. Bring this back and pan it away from the original, keeping both parts fairly even in the mix. The delay isn't so obvious with this short time, but it'll provide a richer sound over the single track alone. Note that if you pan them closer together phase issues will kick in, which probably won't sound so good.

Other track enhancements

Example #1

Make an individual or small group sound like a larger ensemble

Solution

The best solution is during tracking. Have the musician or group perform the same part two or more times onto different tracks. The subtle differences from each performance lend the illusion of a larger group.

A mixdown solution is to use a chorus patch from an effects device. These processors are capable of providing different types of effects, not just reverb. Route the device just we did earlier for reverb. Be sure to experiment with the parameters to tweak the patch for the particular song.

You can also duplicate the track at least once or twice, panning them around a bit, adding some chorus, and perhaps subtle delays. Merely adding a copy, though, won't sound any different; you have to alter it somehow.

Example #2

Make an individual instrument sound bigger in the mix

Solution #1

We used compressors earlier for solving level fluctuation problems, but more often than not these units are loved for the flavoring they bring to a sound. Running a lead vocal through just the right compressor can bring it up front in the mix with a full, rich sound. Back in the 50s and 60s a couple of the favored compressor/limiters was the LA2A and Fairchild

660; they became part of the sound from those records which is hard to get otherwise.

Sometimes this means only slightly compressing a signal through a device, other times you smash the daylights out of it. Each unit responds differently for different results, so experiment.

Solution #2

Set up a signal processing chain on your track consisting of an initial EQ (for cleaning and fixing), a compressor, and then a final tonal shaping EQ. Some engineers will sequence two compressors in a row, finding the result better than hitting one only. There's a reason why DAWs provide more than one insert on each track.

If you're standing there looking at the console patchbay and wondering how to do this with only one channel insert, there's hope. Patch from the channel insert send into the first processor, but then take the output of that into the input of the next processor, and so on, finally bringing the last device output back to the channel insert return. You've now got a chain of processing on the track, just like inserting multiple plugins on your DAW.

Of course there are countless creative possibilities available to the mix engineer. These basic tools can be applied in any number of ways and combinations; only experimentation (and lots of hours on YouTube) will provide a foundation for what techniques can work in various situations.

Audio examples 43, 44, 45: Dry track, chorus, flange

Audio examples 46, 47, 48: Flange, chorus, phaser

Audio examples 49 & 50: Mono organ vs stereo organ with chorus

Audio examples 51 & 52: Dry track vs adding delay

Audio example 53: Vocal without, then with delay plus reverb

Audio example 54: Mix with delay and FX on vocal

Overall things to watch for

Creeping fader levels

Resist pushing faders up repeatedly as the mix develops. What happens is that the guitar player reaches over to turn up her part, then the bass dude pushes his just a bit, then the vocal seems lost and needs to go up. If all the faders are too high, the mix bus will probably be overloaded and you have no more room to adjust balance. Keep faders around unity or lower, especially with higher track counts.

Mix recording level

Always keep one eye on the mix bus meters. If a section of the song really slams the needles, it may overload the mix bus and the 2-track recorder. If your DAW master fader track is hitting the top you'll have very nasty distortion with no room left for mastering. Give yourself some headroom, preferably 6dB or more for digital systems.

Control room monitoring volume

Don't mix at very loud listening levels. Our ears cannot withstand high sound levels for long periods of time. They will tire over time, distorting what we're hearing, and eventually damage will occur. I engineered a session where the producer, a long-time Nashville veteran, needed the volume up so high my ears began distorting after thirty minutes or so. I literally heard distortion and thought at first it must be something overloading on the console or recorder, but instead it was my hearing system actually being overdriven, just like a clipping power amplifier.

The other problem is that humans hear differently at different volume levels. If a song is mixed at high volume, where the ear more easily detects low and high frequencies, you'll tend to not add as much in those bands. So you'll actually hear less bass and highs when playing the song at normal volume on a small system or earbuds. If you mix at low volume, you'll add more lows and highs to make up for what seems missing. This translates into too much of these regions when played back normally. Research the Fletcher-Munson curves for more on this phenomenon.

The solution is to mix around 85dB SPL (sound pressure level). This is a moderately loud setting that provides the best compromise for frequency response (how evenly we hear the entire spectrum) as well as longevity for hearing clearly and not damaging anything. Find an SPL app for your phone and get a feel for what this volume sounds like. Play your mix in the car, on earbuds, and anywhere else to compare how it works in different environments.

Documentation, files, and tape labeling

Just as track sheets document what was recorded on the multitrack, careful notes should be kept during the mixdown session. You never know when you'll have to revisit a project, either to tweak a mix or compile a best-of in ten years. Carefully name all your files, label the tapes and drive folders, note all the settings used during the mix, and include any comments for each mix, such as incomplete, keeper, or "could have been a Grammy if the drummer hadn't puked all over his toms during the fadeout".

Audio examples 55, 56, 57: Too much reverb

Audio example 58: Distorted input on delay unit—level is too high

Audio example 59: Out-of-phase mix

Overall concepts and advanced techniques

Now that you've got a little practice with some basic mixing, let's take it a little farther.

Overall objectives when mixing

- Attempt to capture the magic of what the song is about. Don't lose sight of this while playing with the gear.
- Make it sound like everything's in the same room/space and belongs together.
- Creatively use the sonic space available (stereo field, depth, frequency spectrum).
- Determine what belongs (or doesn't) in the mix: tracks, specific parts, effects, noise.
- Ensure appropriate recording levels in the console (stereo bus, groups) and at the stereo recorder.

Human hearing subjectivity issues

The human hearing system is not linear and does all kinds of strange things to your audio. You need to be aware of how we perceive sound and what that means for your engineering methods.

We translate sounds into a 3-dimensional space that can be manipulated through our processors and controls.

- Width (left-right pan)
- Height (EQ)
- High frequencies are perceived as near the top of the sound.
- Low frequencies are perceived more toward the bottom.

This does not mean that highs float over top of everything and lows run along the floor; this is simply how our brain perceives sound.

- Depth (EQ, reverb/delay)

- Low frequency boosts make a sound appear closer.
- High frequency boosts give you more targeted localization (position in the stereo field). It also makes it seem closer.
- Reverb and delay move things farther back into the mix.

The objective is to manipulate sounds among these variables to make a clear, coherent mix that's appropriate for the style.

Frequency response

As mentioned earlier, the Fletcher-Munson response graph demonstrates that our hearing is not linear from 20-20kHz, but is instead rather uneven throughout the spectrum. This is what allows us to hear the world around us in a very effective way, but the catch is that this response changes depending on volume. Crank up the volume and we tend to hear more bass and treble. So, if you mix really loud, you'll tend to lose the highs and lows during regular playback at home because you heard more of these frequencies in the control room than were really there. Mix at 85dB SPL, occasionally checking the mix at a really low volume to see if anythings sticks out.

Buy quality studio monitors that cover the entire spectrum. Home speakers nearly always push more bass or treble. Also be careful mixing on headphones; stereo imaging is different, you hear subtle things like effects and delays much more, and most of them also have wildly varying frequency response.

Stylistic issues

Different styles and genres of music require different approaches for mixing. I've seen engineers who were well-versed in one genre completely botch a recording that was out of their arena. Listen to different music and discern how it's put together, what it sounds like, what the individual parts are doing. Hip hop is very, very different from blues in many ways, including how it sounds as well as how it's put together. Find a mentor who specializes in a certain style and learn how they work.

Setting up for a mix session

- Turn everything on ahead of time so your gear can warm up. It makes a difference in sound because thermal changes cause slight differences in the operation of electronic components.
- Start setting up the console and processors.
- Patch effects processors via the patchbay so they're ready to use when you're in the thick of mixing.
- Patch outputs into empty I/O line inputs (or aux returns). Pan and assign to the mix bus.
- Pull out the track sheet and label console channels.
- Clean up the tracks by erasing unwanted noises and parts.
- Gate as necessary to clean up tracks.
- Listen to how the gated track sounds in the overall mix; you do not always need complete attenuation.
- Apply low-cut filters to eliminate low-end rumble and leakage.
- Be careful that you don't overlap the frequency range of the instrument on the track, cutting out part of its sound. Set the cut-off frequency accordingly.
- Using filters cleans up the overall mix considerably and reduces clutter and cloudiness. It also allows the monitors to reproduce a cleaner waveform. I also do this a lot for live sound; it gives you a cleaner mix in situations where lots of sound energy is flowing around on stage, potentially muddying what the mics pick up.
- Hi-cut filters can also be useful on some tracks—these attenuate frequencies *above* the desired sounds.

If you're on a DAW, it's all the same thing. Set up any aux tracks and bus sends for reverb and delays, make sure the tracks are labeled, and so on. Built-in templates can be useful as they already have various functions set up and ready to go; you can also build your own over time that matches how you work. Then you don't have to re-create the wheel every mix session.

Starting a mix

- Set up a rough mix to see what it "feels" like.
- This includes fader levels, panning, and perhaps some EQ.
- Keep fader levels conservative—leave enough room to add everything so you don't overload the mix bus.
- Watch your mix bus levels throughout the process.

Which instruments to start with?

Some engineers begin working on rhythm tracks, others on vocals. Practice either way and see which you're comfortable with. It'll be an iterative process, where you set a few things up, add a couple more, then have to adjust the earlier tracks.

Panning "standards"

- Kick, snare, and bass are usually centered.
- High hat can perhaps be off to the side a little.
- Toms and overheads are usually spread out to taste (but not often hard L/R).
- Main vocals are generally in the middle.
- Spread other parts around the primary tracks to make room.
- Pan stereo horn tracks and backup vocals a bit, but not necessarily across the entire field.
- For example, spread a stereo pair of tracks between 9:00 and 11:00 to place one group over towards the left, then place another stereo group over to the right between 1:00 and 3:00.
- We usually pan from an audience perspective looking at the band.

EQ

- Apply corrective EQ as necessary to reduce resonances, edginess, and other issues.
- EQ can also help make space between parts. Guitars and vocals share a similar frequency range, so consider pulling some of the mid-range down a bit in the guitars.
- Apply constructive EQ for desired tone.
- Keep individual channel EQ in perspective to the entire mix— no one is going to hear that guitar track solo'd at home.
- You can only EQ frequencies that are present in that sound. For example, there's no way to add 100Hz to a soprano to make it sound beefier (not that this is a particularly common goal for sopranos).
- "Muddiness" generally falls between 200 and 400Hz depending on the sound. A bad thing.
- "Air" is around 8-10kHz. A good thing.
- Vocal sibilance is around 6-7kHz. Another bad thing.
- 500Hz often features a nasal, honky sound that should be removed with a narrow-band attenuation.
- Low-frequency instruments have a bottom end around 100Hz (+/-), though bass guitars go down in the 30s.

Key frequencies for various instruments

- Kick: body around 80Hz, attack around 2k
- Snare: body 200–300, attack and crispness around 2–4k
- Cymbals: high end shimmer 7k–12k
- Mounted toms: body around 200–300, attack around 4–5k
- Floor toms: body around 100, attack around 5k
- Vocals: intelligibility 2–4k, presence 4–5k, sibilance 6k or so
- In general, many instruments have a body sound in the 200–300 range, with attack and/or articulation transients in the mid-freq range (2–5k).

Compressors

- Helps smooth out a performance so it sits and stays in its place in the mix.
- Adds tonal flavor—different types and models of compressors sound different.
- You can compress individual channels as well as groups, even the main mix at the stereo bus. You get a different sound compressing a group of parts as compared to compressing each individually.
- Level changes can be controlled by a compressor or by moving the fader. Sometimes manually riding a fader sounds more musical than relying on the processor.
- Listen closely for attack/release settings. You might get breathing / pumping effects, which is when the track gets squashed down and then jumps back up to unity gain in a way that you can hear it pumping up and down. Lengthen the release time to prevent the return to unity from sounding so obvious, or reduce the amount of gain reduction overall.

Reverb & effects

- You usually need different reverbs for different tracks or types of parts.
- Make sure to match the RT60 (length of the reverb) with the tempo and rhythm of the song. A long RT60 on a fast-moving dance track will become muddy very quickly.
- Set the wet/dry balance on the effects device output. This controls how much of the original dry sound will be heard along with the effect sound from that device. I usually leave this on 100% wet, relying on the original channels to supply the dry sound—it all gets blended at the mix bus. If you insert a reverb plugin on a DAW track, you'll have to balance this to hear both signals.
- Set delay times to match the tempo and rhythm of the song.
- Calculate delay times: 60,000 / song tempo (bpm) = delay.

- Always listen for phasing when setting short delay times. Phase problems make it sound hollow and thin, so pan the original and delayed signal away from each other to avoid this.
- Consider adding two copies of a track with different delays, then pan differently in the mix. This gives you a very rich sound.

Maintaining a musical balance among your tracks

- Don't press play and throw your feet up on the console during a mix.
- You may have to ride the fader to keep the vocal in the right spot.
- Solos need to blend into the mix, not jump in suddenly. Listen for a smooth volume transition.

How do you end the mix? You'd think it would be so simple.

- Fade outs are easiest, but work to get it sounding musical.
- Sometimes you have to mute tracks, such as all your vocals while the band finishes out the song.
- Use your console's auto-mute function for multiple channels.
- Group tracks so you can fade without grabbing handfuls of faders.

Techniques to try

Re-amp guitar tracks (or even keyboards) to beef up the tone.

- Route the signal from the multitrack output into a guitar amp located in the studio and record to a new track. You can also achieve this through an amp simulator, tube compressor, or anything that can change the sound in an interesting way. Just patch the MT output into the processor and record a track, or run it live during a mixdown. Try the same thing on a track you wouldn't think of, such as vocal, violin, whatever.

Parallel compression

The idea for this is to make a copy of a track, compress it fairly heavily, then blend it into the mix alongside the original. The difference in tone and dynamics between the two tracks sums together in a big way and has been a very popular technique over the years. It's not just for compression—try other processing and treat each copy differently: EQ, delay, panning, compression, effects, etc.

- Send the original tracks (such as drums) uncompressed and routed to the mix bus.
- Create a separate sub-group of these drum tracks and route to a stereo bus or two empty channels. Be sure to match the panning with the original tracks.
- Apply heavy compression to the stereo pair and route to the mix bus. Blend it to taste with the originals; you'll be impressed.

You can do this with any track of the mix, such as vocals, so experiment.

Compressing/processing the mix bus (the entire mix)

- Are you sending it to a mastering engineer? Don't do it yourself —leave it alone.
- Otherwise consider a bus compressor, such as the SSL G-bus compressor. These devices add a gentle compression to the mix that "glues" everything together. This became popular in the 80s and works really well.
- Often (too often) engineers will put a peak limiter on the mix, which adds significant gain on low level dynamics while reducing gain from above. The result is a very limited dynamic range, but louder overall mix.
- Experiment with a mastering plugin; these feature a variety of processing (EQ, compression, aural exciter, stereo width, etc) and have lots of presets to get you started.
- To help ensure your parts are all working together without

sticking out or getting lost, lower your control room monitor level to nearly zero. If anything stands out or disappears you should probably adjust some faders.

Comping tracks in a DAW

Unlike tape recorders, where each take you keep requires a separate track, a DAW can save multiple takes per track. Let's say you had the vocalist record their part three times, hoping for that magical perfect take. If you got one, great. If not, you can find the best pieces from each take and build a final track. This is called compositing, and it's easy to do. In the DAW, set the vocal track to display all the takes above each other. Highlight a region of one take, say verse one, press the solo button, and listen. If it's a good one, click the up arrow (in Pro Tools) to send that section to the main track lane. Do this for each part of the song. Once this is done, clean up the transitions between these by adjusting the waveform clip edges. If you trim the edge of a clip so it overlaps the next, you'll see a cross-fade begin to appear. A very short fade time will smooth abrupt transitions.

Side-chain compression and gating

Notice the side-chain or key input on the compressor or noise gate? Feed an external signal into this and use it to trigger or control the processor.

This opens up all kinds of creative possibilities—here are just a few to think about.

Use the sound from the kick drum to control compression on a bass track

The processor still affects the bass sound, but the control envelope of how the compression actually takes place comes from the kick. Take a copy of the kick track, say with an aux send or bus, and feed this into the sidechain/key input of the bass compressor. A key function is an external "go" switch—when the key input sees a signal, it tells the processor to do its thing. Set the compressor to look for the key input. Play your mix and turn the key input off and on to hear what it does; engineers like this technique because it helps tighten both parts together. It won't be great for everything, so experiment with different mixes.

Here I've inserted a compressor on the bass track. On the kick track I use a bus send and name it "KickCopy". Turn this up and select this bus as the key input on the compressor. To compare how the kick is affecting the bass compression, turn this key input on and off.

Add a low frequency tone to the kick drum

This adds more "oomph" to the bottom end and is common in hip hop recordings. The idea is to use a tone generator and blend it with the kick track. You only want the tone to come through when the kick hits.

- Set the oscillator to whatever frequency best matches the kick sound, say between 40 and 100Hz.
- Patch it into an empty channel, then insert a noise gate. Set the gate to stay closed (threshold).
- Patch a copy of the kick track to the key in of the gate; when the kick hits, this triggers the gate to open, allowing the tone to be heard.

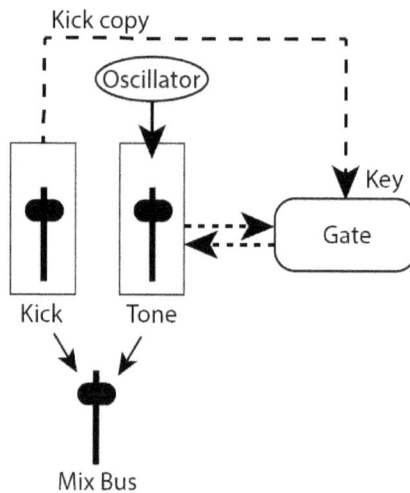

Of course you can do the same thing with sub-harmonic generators, but once you understand the signal flow you can get creative and apply the same technique with a variety of sources. For example, use a percussion track to trigger a sustained synth pad to create an interesting rhythmic pattern.

Ducking

This is used for radio spots and such where you have a background music bed that needs to lower in volume when the voiceover is speaking.

You can do this manually, but why bother when your gear can do it for you?

- Insert a compressor on a music bed being used for your radio ad.
- Take a copy of the voice-over track and key into the music compressor.
- When the voice begins, the compressor automatically reduces the music track, then resumes original volume when the voice track stops.

Of course this process can be used for more musical situations, so be creative.

De-ess a vocal track without a de-esser

The consonant "s" often causes a problem we call sibilance, which is excessive energy around 6 kHz.

- Insert a compressor on the vocal track and set it to key/external mode.
- Feed a copy of this vocal into a separate EQ; boost 6 kHz significantly.
- Take this EQ'd signal and patch into the key input of the compressor inserted on the main vocal channel.

The boosted sibilance frequencies from the external EQ triggers the compressor, thereby reducing sibilance energy without compressing the entire sound. The exaggerated EQ key signal is not heard in the mix; it's simply used as a trigger.

Audio example 60: Kick and bass with bussed parallel compression (out/in)

Audio example 61: Drumset parallel compression (out/in)

Audio examples 62 & 63: Adding tone to kick, first just the kick with tone out/in, then same with entire drum set

Audio example 64: Warming a track by running through a tube processor (out/in)

Once the mix is done

Go take a break. Go eat lunch. Better yet, come back the next day. We tend to lose perspective after listening and working for a long time. It will always sound different later, so avoid a mixing marathon before immediately uploading it to Dropbox.

Once the final mix is recorded on the 2-track or bounced to disk, each song must be individually edited and, in the case of an album, often re-sequenced to match the desired order for the record. Basic editing involves chopping any count-offs and other extraneous noise at the beginning of the track as well as deleting noise and excess silence after the song dies away. We'll cover more of this in Editing & Mastering.

I've presented everything so far assuming you're either working "in the box" in your DAW or on a console in a studio with hardware processors. You can, of course, combine these and get the best of both worlds. Run your DAW tracks individually into the console channels and use your patchbay to insert outboard processors such as compressors. But, take advantage of the DAW's capabilities such as plugins, duplicating tracks, editing, and so on. Even though that vintage hardware compressor in the rack is awesome, there's only one of it, whereas a plugin modeled on that unit can be used multiple times across your tracks. Neither world is better than the other; they just provide different tools and opportunities to be creative.

FIVE
THE TRACKING SESSION

Before we pick up a mic, let's get one thing straight. The main thing that will determine how good of a recording you get is the source itself. It's not the gear, it's the musician. I found this out the first time I stepped in to engineer an album for a national artist. As a recent college grad, I was pretty dumb and inexperienced, and all the local engineers decided to show up to watch the new kid. No pressure there, for sure. I couldn't tell you now what mics I used and how they were set up, but I can say that the professional musicians who played the session made me look really good. You could put a tin can up in the room and it'd sound great with top notch players who know what they're doing (I think it was Folgers). So start with quality sources, learn to hear how they sound, and then develop your miking chops.

Let's first try to understand the signal flow involved in a tracking session. It's pretty simple if you're on a laptop, but on a recording console it's a different ballgame and can get fairly complicated. The concepts are the same, so as we go along just adapt to your situation. Here's the overview of what you're doing:

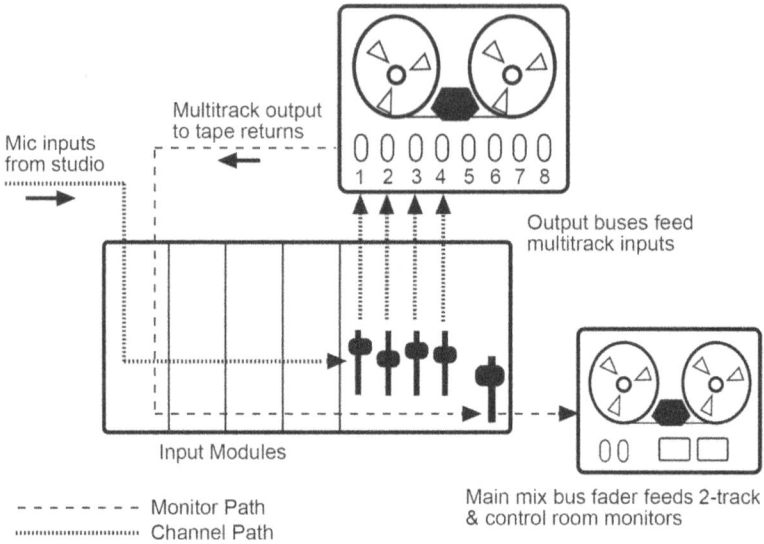

Mic inputs from studio

Multitrack output to tape returns

Output buses feed multitrack inputs

Input Modules

Main mix bus fader feeds 2-track & control room monitors

- - - - - - - - Monitor Path

Channel Path

Tracking on an analog console

Going back to our mixdown discussion, signals start at the multitrack, flow into the console and out to the 2-track, taking a detour to the outboard processing gear along the way. It's pretty much a simple, one-way flow. During tracking, though, signals are running both ways between the console and the multitrack—keeping up with which signal is going where is the trick. Incoming mic signals are coming into the board twice at the same time—from the microphone inputs as well as from the multitrack outputs as they are being recorded. Tracks already recorded are coming into the board, possibly into the same IO modules as that being used by incoming mic signals. Make sense? Of course not, so let's break this down.

It's important to understand what an input/output module does on a recording console. (There are two main types of console designs: inline and split. These will be explained in the console chapter, so for now we'll base our discussion on inline consoles, which are the most common type of analog board you'll run into.) Each channel input has up to three input connectors on the back: mic, line, and multitrack. There are two separate signal paths running down each module, referred to as *channel* and *monitor*. Let's explain these first.

The channel path, also referred to as the mic or recording path, contains the incoming microphone or line-level signal. When you plug a mic in and run it through the board, you are using the channel path for this signal. The channel path signal flows through the IO module, is then routed to an output bus by the assignment matrix (those numbered buttons on each channel), where it is sent to the multitrack recorder. Simple so far.

That signal (along with others which are already recorded) comes back from the recorder into the multitrack returns of the console. The outputs of the multitrack are usually permanently wired to the corresponding channel number on the board, i.e. track one = channel one, etc. These MT return inputs use the monitor path of each IO module, rather than the channel paths. They come in and flow through each module, then are routed to the console's main mix bus, which in turn feeds the monitoring system in the control room (so you can hear everything) as well as any 2-track recorders connected to the mix bus. Not so bad yet.

You can look at this as two separate mixers in one unit. The engineer must set mic levels as they are routed to the multitrack. However, to hear everything, you've got to use the "second" mixer—the monitor paths that get their signals from the multitrack outputs, via the MT returns on each IO module. Both channel and monitor paths have their own level controls.

Now, if you route each incoming microphone to the same numbered track as the channel it's plugged into, things are fairly straightforward. For example, if you've got a kick drum mic coming in mic input 1, you can route it to track 1 on MT. Here you are controlling the recording level as well as the monitor return on the same channel, but using the different physical controls on that module associated with each signal path.

It gets more challenging when you route incoming mics to a different numbered track on the MT. Why would you do this? Say you just recorded a lead vocal on track 17, and the mic is plugged into mic channel 17 on the board. Now you want to record another vocal track, a harmony part. You need to record this to a different track, say 18, but you don't want or need to reconnect the mic into another channel. You've already got all the levels and EQ set for this particular vocalist—no need to reset all this. Just re-assign this channel to track 18 on MT using the assignment matrix. Now, track 17 will still be heard through the console monitor path on channel 17. The new part will still be coming in the board on channel path of module 17, but you will use the monitor path on module 18 to hear it.

Thus two different signals are simultaneously running through module 17. The first vocal track is coming in on monitor 17, the new part coming in channel 17 being routed to track 18 on MT, then returning into monitor 18 on the console. Make sense? It will once you actually try it yourself, though it takes awhile to become really comfortable, especially during a hectic session. Here is the summary of where a microphone signal goes during a recording take:

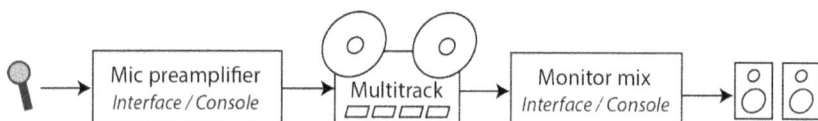

The overall issue to remember is to keep channel and monitor paths clear and distinct in your mind. They are separate signals, even if they happen to be flowing through the same numbered channel module on the console. Certain functions are always done to the channel path signal, others to the monitor path, and sometimes it depends on the situation. So, the mic signal always flows through the channel path on its way to an output bus that sends it to the multitrack recorder. These signals always return through monitor paths on the console that feed the mix bus (and therefore your control room speakers). Now, do you want to EQ a mic signal so it gets recorded that way? Set the EQ into the channel path where the mic is plugged in. Do you want to record a snare drum with compression? Insert the compressor into the channel path of the mic signal. Cue mixes are almost always sent from the monitor paths (multitrack returns), so turn up your cue aux sends from the modules corresponding to the tracks on multitrack (not the mic channels)—then make sure you source these to monitor path. It'll take awhile for this to become clear, but focus on these principles while you practice and you'll get it. Some console manufacturers helpfully distinguish the two signal paths visually in some way, such as shading the background color differently.

Tracking without a console

If you don't have a console it's much more straightforward. The mic inputs on the audio interface will be assigned to individual tracks in the DAW. You can change these as needed by simply selecting a different mic input directly from a track. The channel path mic signal we've been talking about runs from the audio interface and is recorded straight to a track. You don't have any DAW control over the mic signal before it gets recorded, just the preamp level on the interface.

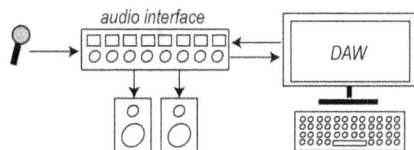

What you see in the DAW tracks are monitor path, including what may have been recorded earlier. So this operates just like a regular console, including faders, pan pots, aux sends, and inserts for EQ, compressors, etc. All tracks route to a stereo output to feed your monitors.

Getting the equipment ready: Control room setup

You need to arrive much earlier than the clients so you can get everything ready for the session. Here's what you need to do:

The room

Turn on all equipment as soon as you arrive. As electronic equipment warms up the calibration and operation of the internal parts adjust. What this means is that the gear will sound a bit different after it's been on awhile. Many studios leave their main devices such as consoles and tape machines on all the time.

Check to make sure everything is working. You can't afford to wait until the session actually begins and then discover the console has flipped a logic board or something. There will be plenty to worry with when everyone arrives—do what you can to reduce session delays and tense emotions.

The console

For now, follow these steps in order. Eventually you'll learn which ones really matter, like not turning on phantom power after everything else is routed (it makes a loud pop that could damage speakers).

The individual channels (left side of the console)

Select the signal source

For each input channel you will use the microphone input as the source. All consoles have a mic/line select switch, so make sure it's set to mic.

Phantom power

If it's a condenser microphone, turn on the phantom power switch. Don't do this later after everything is up and running or it'll pop loudly in the monitors.

Assign to output bus

Each incoming signal must be routed to some output. Since we are tracking microphone signals to the multitrack recorder, we need to assign each channel to a particular output bus, which in turn corresponds to a track on the MT. Consoles have several numbered buttons either at the top or bottom of each module; this is called the assignment matrix. Pushing "12" will send that mic signal to track 12 on the multitrack.

Often consoles will save space by using one button for two output buses, such as 1-2, 3-4, etc. There will either be a *shift* switch to select the second option, or you pan all the way to one side. So, if you want a mic to go only to track one, simply push the 1-2 button and pan that channel all the way to the left. Another mic coming in a different channel can be recorded separately onto track 2 by pushing the same 1-2 button, but panning instead to the right.

Channel fader level

Each console channel will have either two faders, or a single fader and level pot. These provide level control over channel and monitor path, so for a mic signal push the channel fader up to unity (0).

Output bus level to multitrack

Once the signal is assigned to an output bus, that bus level needs to be turned up to set the recording level to multitrack. Multitrack buses are controlled by a row of pots or faders in the master section of the console.

Once you select an output bus in the assignment matrix on a channel module, go to that respective output bus fader and turn it up to unity (shaded area). So, if a mic signal is coming in channel 1 and being assigned to track 3, turn up output bus fader #3 in the master section. This fader is directly wired to its own input on the multitrack recorder.

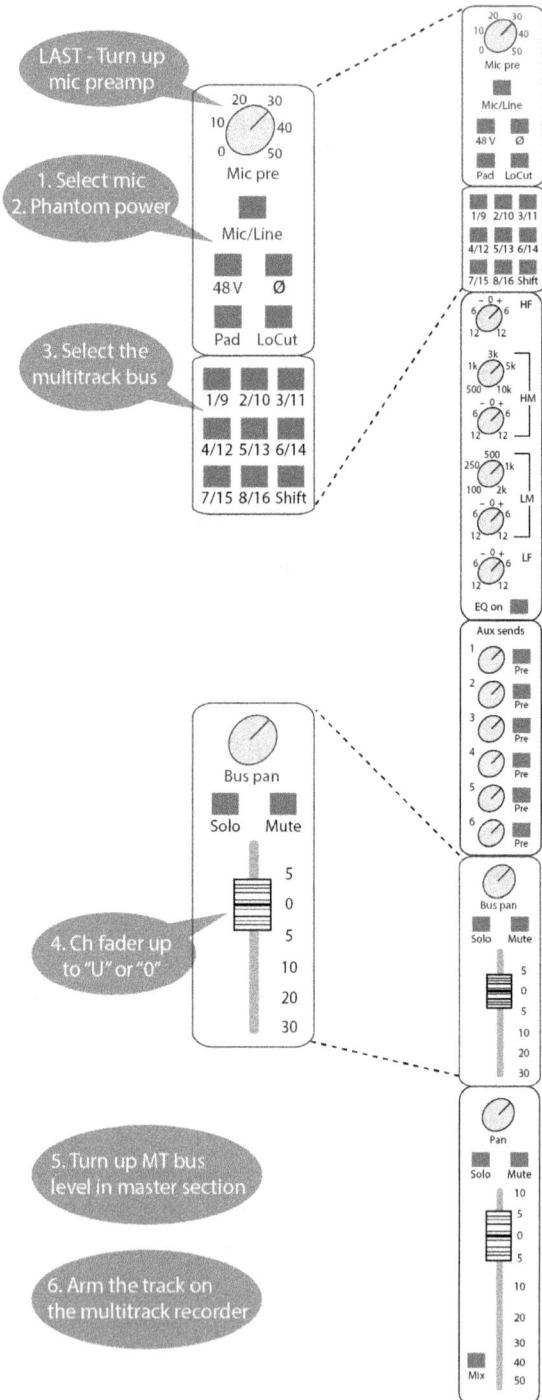

LAST - Turn up mic preamp

20 30
10 40
0 50
Mic pre

Mic/Line

48 V Ø

Pad LoCut

1. Select mic
2. Phantom power

3. Select the multitrack bus

1/9 2/10 3/11

4/12 5/13 6/14

7/15 8/16 Shift

Bus pan

Solo Mute

5
0
5
10
20
30

4. Ch fader up to "U" or "0"

5. Turn up MT bus level in master section

6. Arm the track on the multitrack recorder

20 30
10 40
0 50
Mic pre

Mic/Line

48 V Ø

Pad LoCut

1/9 2/10 3/11

4/12 5/13 6/14

7/15 8/16 Shift

- 0 +
6 6
12 12
HF

3k
1k 5k
500 10k
HM
- 0 +
6 6
12 12

500
250 1k
100 2k
LM
- 0 +
6 6
12 12

- 0 +
6 6
12 12
LF

EQ on

Aux sends
1 Pre
2 Pre
3 Pre
4 Pre
5 Pre
6 Pre

Bus pan

Solo Mute

5
0
5
10
20
30

Pan

Solo Mute

10
5
0
5
10
20
30
40
50
Mix

Record ready on multitrack

Arm the track on the multitrack recorder that corresponds to the output bus you selected. Tape machines should be in either input or sync mode. For a DAW, just click the record ready button on the track you're getting ready to record to. After arming the track you'll be able to see signal on those meters once you turn up your levels.

Monitoring source and volume

This was described in the mixing chapter and is the same here. Push up the main mix bus fader to unity, select main mix as your control room monitor source, and turn up the volume pot a bit—be careful for now. This will then monitor everything coming through the monitor paths, which should include everything on the multitrack if you've set it up correctly.

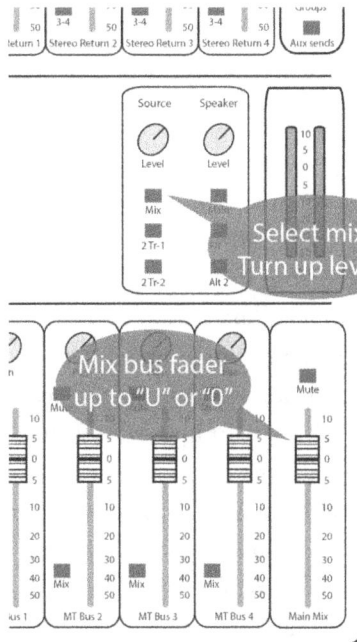

Monitor path / MT return

Remember, we do all our listening via the monitor path returns from the multitrack recorder. After the mic signals have been routed through the IO module to the output buses and to the multitrack, those signals are immediately routed back into the MT returns of the console. Use these channels for listening to what you are recording.

On each console channel that corresponds to a track you're recording to (or already recorded earlier), turn up the monitor level control. This is either the main fader at the bottom of each channel (different from the channel path fader) or a pot next to the pan control. Turn it up to unity. Some consoles have a switch (in the assignment matrix or near the monitor fader) that assigns the monitor path signal to the mix bus; others automatically route it.

Mic preamplifier (trim)

The last step is to set your recording levels to the multitrack. At this point you may only be seeing a very low signal on your meters, if anything at all. That's because one of the most crucial stages in the chain has not been turned up yet—the microphone preamplifier. As the musician is making noise into the mic, slowly turn this control up and watch your meters carefully. Leave all the other level controls where we showed you, such as the output bus faders. Use the mic trim pot to fine-tune your levels to multitrack. The reason this control is so important is that it's the mic pre that contributes greatly to sound quality. Many studios spend thousands of dollars buying specific types of mic preamplifiers—they all have their own unique sound, just like microphones. The job of a mic pre is to take the very low electrical signal level of a mic and amplify it to the much higher line-level used by the console and other equipment. This much amplification really affects the sound, so the pre must be well-designed (many of them are not, so be aware of this when you decide to buy that cool-looking-half-price model from your favorite mail-order catalog).

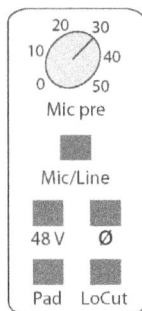

You should be getting levels on both the console and multitrack recorder meters. Make sure the musician is providing a decent sound for you—many untrained studio musicians will simply talk into the mic, which is much lower in volume than when they actually sing or play. Have them do what they will be performing during the song—and remember that when they actually begin recording it always gets louder. They get excited when it's for real—just getting sounds to multitrack is not much to get the adrenaline pumping.

Make sure you're not using your ears to set this level. The monitoring volume is indirectly related to what you put on multitrack. You might be barely hearing anything, yet overdriving the recorder or distorting the input to the board. Always watch the meters on both the console and the multitrack.

Summary of console setup

In order, here are the steps we've described above.

To get signal into the board and to the multitrack recorder

- Select mic input as the signal source (mic input switch).
- Assign this signal to an output bus (assignment matrix).
- Turn up output bus level to the multitrack (output fader or pot).
- Arm tracks on the multitrack recorder.

To hear what you're doing

- Turn up the main mix fader.
- Select main mix as the control room monitor source and set the volume.
- Turn up the monitor path fader on each multitrack return channel.

To set final recording level to multitrack

- Turn up mic preamp while the musician is playing or singing.

Setting mic levels without a console

Arm the DAW track and turn up the mic preamp on the interface. Everything is automatically routed to the mix bus and your monitors. Use the interface preamp for fine-tuning the recording level. That's it.

Cue mix

Now that you've got levels to the recorder and can hear them in the control room monitors, a separate mix of these tracks needs to be sent to the headphones in the studio. This is called the cue mix, and this is what allows the musicians to hear what is already on multitrack as well as the new parts they are currently recording.

Use an aux send from each channel to send out to the studio headphone system. If you recall, we used auxiliary sends during mixdown to send copies of certain tracks to the outboard processing gear, such as for reverb. Aux sends don't necessarily mean reverb—they are merely additional outputs that take a copy of that channel's signal and send it wherever you want.

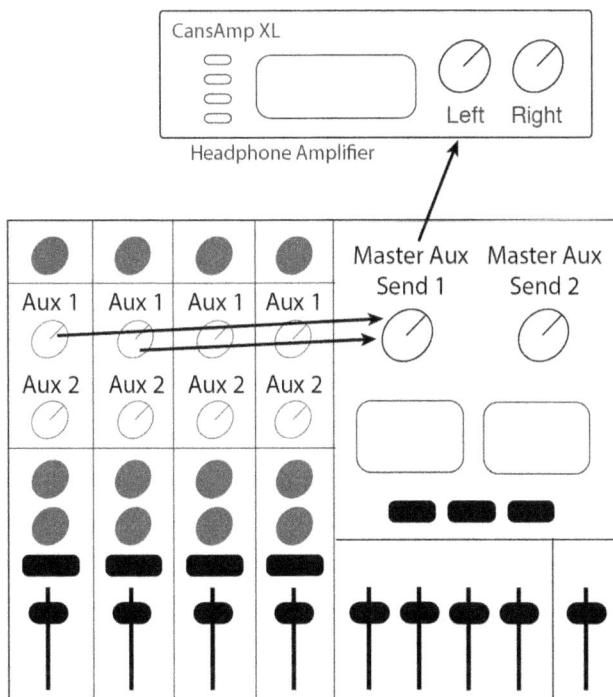

Aux sends work well for cue mixes because you basically create an entirely separate mixer. For example, many studios will permanently connect aux 1 & 2 from the console to their headphone amplifier. Since there is an aux 1 & 2 on each channel, you can send any (or all) of your tracks to either (or both) of these outputs. Why two auxes? You can set up your studio to have two entirely separate cue mixes, or you can have one stereo mix, aux 1 feeding the left side of the headphones and aux 2 feeding the right side. Many recording consoles conveniently provide stereo aux sends, so you can use one send and pan accordingly—much easier.

One last important step—make sure you push the *pre* button beside the aux send control on each channel. Simply put, an aux send takes a copy of the signal flowing through a channel and sends it wherever you want. Now, the issue is where exactly in the channel does it come from. A normal aux send setting comes after the main channel fader, so any changes in this fader level will also affect how much goes out the aux send. During a tracking session, however, the musicians need their own mix of the tracks, and they don't want this mix to change while the engineer and producer move faders around (which is inevitable). So, we need to source the aux signals *before* the channel faders, which is what the pre-fader button does. Bottom line—every time you are tracking with a cue mix, automatically select all cue aux sends as pre-fader.

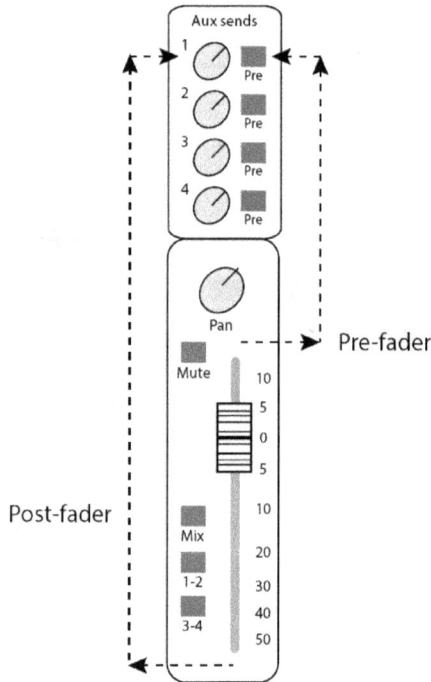

The master aux outputs are then connected to a headphone amplifier, which in turn feeds the headphones in the studio. (All speakers, including headphones, require amplified signals to work.) Some mixers provide a built-in headphone amplifier.

For cue setup without a console, all of this works the same. Turn up sends on individual tracks, create an aux track, set your cue bus send as its source, then route the output to a pair of interface jacks different from the main mix that's feeding your speakers. Connect these to your headphone amp. The interface must have more than two output channels to do this; if not you'll have to send the main monitor mix using a headphone splitter.

Once you become proficient at setting up the console for a tracking session you should get into the habit of establishing the cue mix as soon as possible. The musicians are already in the studio attempting to play into the mics and hear themselves—it's very frustrating for them if it takes forever to hear anything in the cans (headphones). It's better to go ahead and turn up the auxes you will use on all channels and get the cue ready for signal—then when you turn up each mic level the cue mix will already be set.

One last word of advice here. All consoles have the feature of selecting the aux outputs in the control room monitors. Remember how you selected the main mix as your listening source? You can also solo or select the master aux outputs to hear their respective mix of the signals. Listen to this periodically so you can hear the same blend that the musicians are getting. It's never the same as that in your control room mix, so sympathize with them and make sure you understand what they're getting in the other room. Most professional session players would tell me their main complaint is about engineers who don't seem to care what's coming through the cue mix—since they don't have to hear it themselves they're oblivious to what the musicians are dealing with in the studio.

The multitrack recorder

If you're using an analog machine, clean the entire tape path thoroughly and load a reel of tape. Make sure it's a new reel, or at least make sure it's not the album tracks you just recorded the day before (there is no *undo* in analog). You should splice leader tape onto each end of the tape; otherwise you will be using recording tape merely to thread onto the reels. Save the recording tape for music. If this tape has already been used and you intend to erase over it, go ahead and place all tracks into record-ready and erase the entire tape before the session. This will prevent unwanted sounds from popping up during the new project. Once you're finished, place the tape machine into input mode so it's ready to accept a signal source from the console.

For DAWs, create a new file with the following settings: WAV or BWAV, 44.1k sample rate (or higher if you want higher resolution and your system can handle it), and 24-bit. Add some tracks and place them into record-ready so you can see signal level on the meters. Name your tracks as you go, before actually recording anything, so your growing audio clip list follows a meaningful naming convention.

Signal processing

If you are planning to record any tracks with signal processing such as compression or gating, patch these devices into the appropriate input channels on the console. Note that you must patch into the channel that the mic or direct box is currently plugged into during the performance— not necessarily the track # of where it's going on multitrack.

In other words, you might have a vocal mic plugged into input channel #1 on your console (from the studio), but need to record it to track 15 on the multitrack (a common occurrence). If you want to compress or do something to the signal on the way to the multitrack, meaning record it with the processed changes, then the signal processor must be inserted into the signal path *before* the multitrack. In this case that would be the channel path of channel 1.

Now adjust your processor settings as desired; you can't change it after the recording.

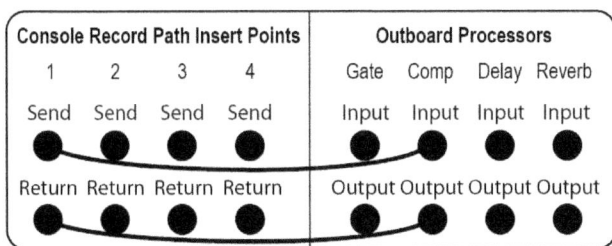

Console Record Path Insert Points				Outboard Processors			
1	2	3	4	Gate	Comp	Delay	Reverb
Send	Send	Send	Send	Input	Input	Input	Input
●	●	●	●	●	●	●	●
Return	Return	Return	Return	Output	Output	Output	Output
●	●	●	●	●	●	●	●

Studio setup

Mics, cables, and stands

Before the musicians arrive, arrange mic stands and cables around the room according to your studio layout chart. This plan should have been devised during the initial pre-production meetings between yourself and the clients, and it determines where instruments will be located. Allow plenty of extra cable at each stand in case you need to move it. Coil it neatly at the base, and make the runs across the floor as neat and straight as possible. Avoid scattering cable all over the floor to avoid people walking on it; bundle them together as possible to keep the floor space free of clutter. Cables are expensive and fragile—discourage people from walking on them as this will damage them over time.

Remember to screw the boom into the microphone clip, rather than trying to spin the mic and dropping it on the floor. Hold the mic securely while inserting the cable connector; listen for the click to indicate it's in place. Wrap the cable loosely around the stand to keep it from hanging out and tripping anyone. And don't leave a mic on the floor—you just might be the moron who steps on it. Leave it in the case until you're ready to set it up.

With an assistant, check each mic to make sure it's working. Turn up the preamplifier slowly until you get a signal. Replace any defective cables or mics before the musicians arrive since there will be plenty to worry with once they are there.

Don't touch or move mics around until the channels are muted and the preamps turned all the way down on the console. This especially includes plugging and unplugging mics. You can do serious damage, including to your musician's ears. Warn everyone not to touch the mics and to let the engineers handle any adjustments.

Documentation

Create a mic input chart which indicates which microphones should connect to specific inputs on the studio patch panel. Without this chart it can be frustrating remembering which mic goes where during a complex session. Get organized first before you start running cable across the room.

Remember, these input numbers are not necessarily the track numbers on the MT. Once the mic signals come in the IO modules they are routed individually to whichever track you want. In other words, a vocal mic coming in on channel 7 could be routed to track 15 on MT. You would set up your monitor levels for this part on monitor path channel 15 on the console as described earlier. If you can keep these different numbers straight you're more than halfway there (honest).

Cue system

Get the headphone system for the studio ready. This involves connecting headphone cue boxes to a designated jack on a panel in the studio. This output jack from the headphone amplifier might be daisy-chained into multiple headphone stations around the room, or you may have to connect each headphone mixer to a different jack on the panel. Keep each individual box close to the mic stand within easy reach of the musician. Connect a pair of headphones to each box and hang them on a music stand. Do not hang across the microphone itself. Run something through the cue system once the console is set up to make sure the cue is working okay.

Where do the musicians go?

This depends on the situation. To reduce leakage between mics you will be tempted to spread everyone out as far as possible, even putting some people in isolation booths. Leakage is when you have an instrument being picked up by another instrument's microphone. The sound of the wrong instrument in the mic doesn't sound very good since the mic isn't properly placed for the other instrument. This will be covered more in the microphone chapter. From an engineer's standpoint leakage is often not desired, depending on the style of music (a jazz combo sounds natural when everything is filling the space, for example). However, separating musicians tends to conflict with the musical needs of the group. It's much more difficult to play as a tight ensemble when everyone is separated. So instead of assigning people to different rooms, or overdubbing each part separately, consider spreading them around the room and putting gobos in between. The farther a mic is from a sound source the less it will pick it up, though condenser mics are far more sensitive than dynamics. Increase the distance as much as possible with mics facing away from other sources. A gobo is a portable barrier that can reduce sound transmission, though it won't do anything for bass frequencies.

Work with the clients to determine the most satisfying arrangement. Above all else, make sure the musicians are comfortable. If they're not, then you'll get a less than optimum performance and the great sound won't mean anything. The music always comes first.

Tracking procedure

Basic tracks

Once everything is ready you can lay down the first basic track. Basic tracks are the initial parts recorded that serve as a foundation for the song, such as the rhythm section.

On the multitrack recorder, arm the tracks that you will be recording during this pass. Get the musicians quiet in the studio, press record-play on the multitrack, and signal the musicians. This is usually done through the console's talkback mic, which is fed into the cue system headphones.

As the band is playing, don't just sit back and check your Instagram feed. Watch recording levels to make sure nothing goes over zero; if so, go back after the take and audition that point to check for distortion. On the track sheet, notate which tracks are being recorded and indicate time locations for cue points in the song, such as first verse, first chorus, second verse, second chorus, bridge, etc. This is easily done in a DAW with memory locations; in Pro Tools press *enter* and then name that location. You can even do this while the recording is in progress. These are crucial for quickly finding any section of the song later to fix a part or add a new track.

Fixing and overdubbing

Once the first pass is complete, immediately go back to the beginning of the recording or to a section that needs to be fixed due to a mistake during the performance. Professional musicians will immediately tell you where they need to fix a note or phrase—if you are a musician you should already have heard and noted these problems. Don't just dumbly wait to be told what needs to be done. Try to anticipate what the producer and musicians will need to work on. Have copies of the music charts or scores so you can follow along—it will speed up the session and help you work more closely with the others.

Disarm the tracks except for the part that needs correcting. To fix a bad spot, referred to as *punching in*, rewind before the problem location, play the MT and have the musician play along. When you get to the spot, find an opening in the performance and quickly hit the record button. The recorder will continue playing, but will now be recording the new performance on that track. When you pass the problem area, find an opening and hit the play or stop button to take the track out of record mode. Go back and listen to make sure the punch is smooth. You may need to take a minute before recording to find a decent place to punch in and out—it can be difficult to get it right, but it gets easier with practice and experience. It's far easier on a DAW as you can set punch in and out points, set a pre-roll for it to play a few bars before the punch, and then if something goes wrong simply *undo* and adjust the in/out points.

Once the basic tracks have been fixed, you can either go on and record basics for the next song or begin overdubbing on the first tune. If you are changing mic setups for the next recording, make sure you bring down all mic preamp and channel fader levels. Anytime you move or plug/unplug a mic you send a serious jolt through the system which can damage something.

Remember how we described the console monitor mix setup? Since all your monitoring has been established from the multitrack returns, and not the mic inputs, the basic tracks you've just completed will still be heard through those monitor channels—even though the microphones are no longer connected. To overdub a new track, simply route the new mic input to an unused track and turn up the monitor levels on that console channel (don't forget the cue aux sends).

Confused yet?

It takes awhile to get used to all the details involved in tracking—that's why we started with mixing. Just keep it simple as you practice and follow these steps carefully so you gradually get the hang of it. Once you feel comfortable with the buttons, faders, and numbers, spend more of your energies listening to the sounds you are getting from the mics and equipment. The main priority for getting high quality sounds is placing an instrument in a good room, putting the right mic in the right place, and having a great musician play. However, you cannot concentrate on this if you're still figuring out which buttons to push and which to avoid. The next section might best be read after you've gotten the hang of it and are ready to get down to business—we're going to explain some of the miking and recording issues you'll run into and how to handle them.

Getting great sounds from your mics

Just as we did in the mixdown chapter, we're going to point out several situations that occur during tracking and provide some possible solutions. Learn to use your ear and experiment; once you're comfortable with the equipment and procedures you can pay more attention to the more important task of getting good sounds. Some of these will sound very similar as in the mix chapter—they are, and this is the time to make sure these problems don't get recorded to begin with, requiring you to fix them during mixing.

The snare doesn't sound good

Many times you'll place a microphone on an instrument, turn it up in the control room, and wonder what happened. You look to make sure that it is indeed a snare drum you're working with, but it doesn't sound right.

Solution

Don't reach for the EQ. Go into the room and listen to the drum itself. Get a feel for what it's supposed to sound like so you know what you are shooting for. Here are a few things to try:

Move the mic around. Even a few inches in either direction can make a big difference, especially in terms of how close the mic is. Acoustic instruments need space and time for their sound to develop, so if you put a mic an inch away from a drum head you're never going to get the fullness of what the drum actually sounds like. Pull it back a few inches and see if that helps.

Use a different mic. All mics are not created equal, and that's intentional. Each mic has its own characteristic sound and particular way it picks up different sound sources, so experiment. Over time you'll build your own repertoire of what mics you like for certain instruments and voices.

Work on the instrument itself. A good musician knows how to fine-tune their instrument for the best possible sound. Some have no clue, so the more you know about different instruments the more you can fix things for them.

Move the instrument in the room. Many rooms have different sonic characteristics throughout the space—try another location and see what happens.

The main issue here is to avoid the common reaction of using EQ to fix bad mic choice and placement. Your goal should be getting as clean and perfect a sound as possible into the mic itself—then if needed use some EQ or other processing to help out.

I can't get enough level on the mic

Once you set the console and recorder and begin turning up the mic preamp, you may not get enough signal on the meters, even if you go all the way with the pot. There are a couple of reasons this can happen.

The musician is not playing or singing at full volume. Often a musician will only talk into the mic, or lightly play their instrument since they're not in the full swing of the tune. This is a much lower sound level than you'll get during the recording, so try to have them play normal—ask them to rehearse their parts or something.

Dynamic mics have a lower signal output than condenser mics. The microphone chapter will explain these terms, but for now just remember that some mics aren't as "loud" as others. If you put an AKG 414 on an acoustic guitar, then replace it with a Shure SM57, you'll have to crank up the 57 higher to get anything through the console. (The reverse is also important: if you have a dynamic mic on a source, then replace it with a condenser without changing levels, you'll likely get a nasty surprise as your speaker cone shoots across the room...well, maybe something like that.)

The mic may need to be closer. Many sources such as guitars with nylon strings and quiet singers don't put out a lot of sound, so you need to move the mic closer. It's better to get closer than having to crank all your gain controls wide open.

If you still cannot get enough level to the multitrack, and you're certain it's not your equipment (you've successfully gotten other mics to work), then you can use the output bus fader as another level control. We said earlier that the output buses should be set at unity and left alone, but if you need extra amplification that the mic pre cannot provide, then go ahead and turn these faders up. Some consoles actually have two different level controls in addition to the mic pre, so balance these to help (don't turn one way up and leave the other way down).

It sounds muddy and boomy

There are a couple reasons that your sounds are muddy or boomy on the low end.

Explanation #1

With certain types of microphones, close placement to the sound source results in a low-frequency boost called proximity effect. This occurs with directional mics such as cardioid (uni-directional), hyper-cardioid, super-cardioid, and bi-directional. Omni-directional mics are not affected by this.

Example

The acoustic guitar sounds unnaturally boomy when close-miked to the instrument.

Solution #1

Move the microphone a little farther away from the source. This will attenuate (reduce) the low-frequency boost.

Audio example 65: Moving mic back from source to reduce boominess

Solution #2

Make sure the mic isn't pointed directly at the sound hole on a string instrument.

Solution #3

Switch to an omni-directional microphone since they do not suffer from proximity effect.

Solution #4

Turn on the low-cut filter on the microphone (if it has one) or on the console channel the mic is plugged into. This will sharply attenuate all frequencies below a set point, usually around 75 or 80Hz. Of course, for instruments that have low-frequencies, such as upright bass and kick drum, you will begin to lose the sound of the instrument itself, so be careful. By the way, I usually wait and apply filters during mixing, when I can spend more time dialing it in to match the mix.

Audio example 66: Reducing boominess with low-cut filter

Explanation #2

All acoustic sound sources have a resonant frequency range, which is a region of frequencies with higher amplitude than the rest of the sound. These are usually in the low-mid range and will make the sound cloudy or muddy. If this isn't cleaned up on individual tracks, the combined effect will result in an overall muddy sounding mix.

Example

Muddy sound in a kick drum or upright bass.

Solution

Use a parametric EQ to find the offending frequency range.

A parametric EQ provides an extra control that allows sweeping around the frequency spectrum to find specific frequencies to boost or cut, as opposed to a graphic EQ which is set to a fixed frequency. All recording consoles have parametric EQ and DAWs feature a variety of EQ plugins to choose from. For each band (high, high-mid, low-mid, low) there will be two controls: one to boost or cut the signal level, another to select the desired frequency.

How to find the resonance

Listen to only the kick drum mic channel by soloing the monitor level fader on the MT return channel. Turn on the EQ. Turn up the gain for the low-mid EQ at least 6dB or so. Now rotate the frequency select control adjacent to it and listen for the changing sound of the frequencies as you move up and down the scale. Turn back and forth until you hear a region which gets louder and more muddy. Now reduce the boost control back to a negative number, at least -3dB, maybe as low as -9. The more you cut the more of the overall sound will be taken out, possibly thinning the sound too much.

Most parametric EQs provide one additional control that allows you to narrow the region of frequencies being cut (bandwidth, or Q). By narrowing the bandwidth of affected frequencies, less of the overall sound will change, allowing you to attenuate (cut) only the offending frequencies.

This same procedure can be used to isolate other EQ problems, such as harsh hi-mid tones.

Audio example 67: EQing an acoustic guitar to clean up muddiness and improve clarity

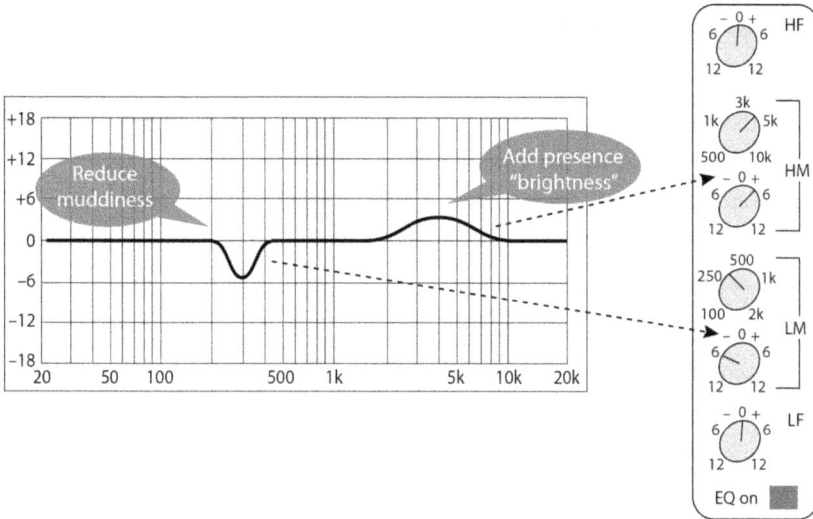

You hear everything in one microphone

If more than one instrument is in the studio at one time during a tracking session, individual microphones might pick up other sounds besides the instruments they are assigned to. This is called leakage. The technical solution is to try isolating the sound sources from each other through a variety of ways, though this often causes conflict with the musicality of the performance. Groups play best when they play together, in the same room, and can see and hear each other. Isolation techniques work against this principle, so you need to balance these issues when deciding on a studio layout.

Example

Drum sounds leak into the piano mics

Solutions

Separate the musicians into two different rooms, such as moving the drums into an isolation booth. Try to maintain visual communication between the musicians.

Keep both instruments in the same room, but put gobos between them. Gobos are movable partitions designed to help block sound. The problem here is that drums are so loud it's nearly impossible to contain the sound within that room. Gobos can be more effective with acoustic guitars, vocals, etc.

Overdub one of the parts, most likely the piano track, later. This completely eliminates the leakage, but also destroys any real-time interaction between the musicians. For some tunes this is not critical; for others it can kill the recording.

Another option with a grand piano is to tape PZM mics (they look like flat plates—see the mic chapter) underneath the piano lid, then close the lid all the way. Cover the piano with a heavy cover and other blankets if needed. For some projects, the resulting sound works fine and the leakage is greatly reduced. It's a very different sound, though, than using other types of microphones.

Watch your recording levels

A musician's dynamic range can vary a lot between soft and loud passages. This is especially true for vocal parts, where the singer may whisper a verse, then scream through the chorus. The problem is setting the recording level high enough to hear the whispers, yet prevent distortion during the screaming. With modern 24-bit recording, this isn't much of a problem during tracking; simply reduce the mic preamplifier so the screams don't overload the track and use a compressor during mixdown to make it fit into the mix. But if you want to better control a widely fluctuating performance, try compressing the mic signal on the way down.

Solution

Insert a compressor into the mic path to smooth out the variation in levels.

A compressor acts like an automatic volume control. It reduces the level on audio signals which get too high and may cause problems for the overall mix. How much compression to use, and what threshold to set, depends entirely on the situation. You can set a higher threshold to control only the peaks, without noticeably affecting the sound itself. If you lower the threshold, the compressor begins working on the overall signal, which begins to change the sound. There's no correct answer here—do whatever sounds good.

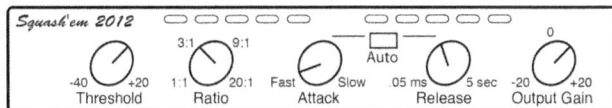

How do you do this?

Patch the channel insert send for the mic channel to the input of a compressor through the patchbay. Then patch the output of the compressor back into the insert return for the same channel. Remember each channel of a compressor only affects the single channel it's inserted into. If you are compressing or gating a stereo pair of tracks then you must connect a compressor channel for each track.

Console Record Path Insert Points				Outboard Processors			
1	2	3	4	Gate	Comp	Delay	Reverb
Send	Send	Send	Send	Input	Input	Input	Input
●	●	●	●	●	●	●	●
Return	Return	Return	Return	Output	Output	Output	Output
●	●	●	●	●	●	●	●

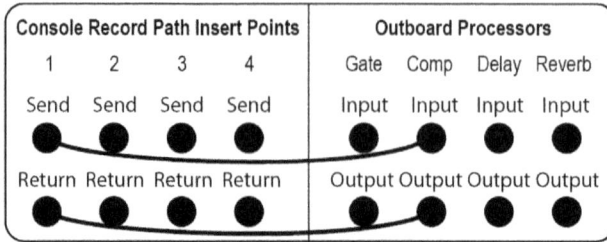

If you're running a DAW with no console, keep in mind that while you can insert a plugin on the recorded track, it actually doesn't affect the signal as it's being recorded. DAW tracks are post-recorder, meaning it's all in the monitor path. No big deal, actually, as the end result is the same. Before the days of DAWs and plugins, however, engineers only had whatever hardware was in the rack. We had to be proactive in dealing with level issues during tracking so the limited number of processors could be used during mixdown for more creative purposes.

The threshold setting indicates which levels above it will be compressed (reduced). Therefore signals hotter than the threshold setting will be compressed; the amount of compression is determined by the ratio control. The numeric ratio markings signify the amount of incoming signal vs output signal. In other words, a 3:1 ratio dictates that for every 3dB of signal going into the compressor (and above threshold), only 1dB of that signal will come out. Therefore, a higher ratio such as 5:1 will provide a greater amount of compression.

If you only need to knock off a few high peaks, but don't want to affect the entire dynamic range, set a higher threshold (zero or above) and higher ratio (5:1 or greater). If the whole signal needs to be compressed to tighten the entire range, such as for a rhythm guitar or bass guitar, then set a lower threshold so more of the original signal will be compressed.

Watch the gain reduction meters and listen how it changes with different threshold and ratio settings. Learning compressors takes lots of practice and experimentation. Go ahead and play around with the controls to see how these devices affect sounds.

The vocalist isn't quite in tune

When a singer wears headphones the acoustic isolation from the room prevents them from hearing the natural sound of their voice.

Solution

Have them pull the headphones off one ear so they're hearing the cue on one side only.

What's that sound in the background?

Any background noise that's not part of the music should be reduced as much as possible, preferably during tracking. Air conditioning vents are a common problem, along with other building noises, traffic outside, birds nesting (it happens), and so on. Keep mics away from vents as much as possible, or even consider turning the AC off during each take. Larger vents are quieter, but I doubt you're going to replace all your ductwork.

Anything else to make it sound better?

Along with having a great source in a good sounding room, with the right mic in place, the best thing you can do is acquire high quality mic preamps. Sometimes the one in the console is perfect for the moment, but having an array of different preamps gives you options. Each preamp offers a different sound in the way it handles mic signals. It's also a good idea to keep the mic path as short as possible, so consider plugging the mic directly into a mic preamp, then run from this straight to your multi-track recorder. Don't use the console for channel/mic path at all—only for monitor path returns so you can monitor the mix and send a cue mix to the studio.

Although you can experiment with processing mic signals on the way down, such as EQ, compression, etc, there's no real need for that anymore. Just get a clean signal during tracking and then spruce it up during the mix.

Having problems getting signal?

Follow these steps to see what's wrong

Check your console monitor setup

- Main mix bus selected (2-mix, L-R mix, stereo mix)?
- Main monitors selected?
- Control room monitor volume up?
- Is something solo'd? Check all I/O modules, aux sends/returns.

Check your multitrack recorder

- Correct mode: input, sync?
- Track in record ready?
- DAW track source match the audio interface input?

Check your console I/O module setup

- Mic/Line switch correct?
- Phantom power on (for condenser mics)?
- Mic preamp turned up?
- Channel fader up?
- Bus assignment selected?
- Output bus level up?

If you're using a dynamic mic, remember that your source may not be playing or singing very loudly for a sound check. You may just need more gain—but be careful that everything is set correctly first.

Check mic and connections in the studio

- Turn down mic preamp & mute the channel first.

- Check switches on the mic (pads, polar patterns).
- Make sure mic cable is plugged securely into the mic.
- Verify mic cable is plugged securely into correct mic input.

Still no signal? Go through each of these one at a time.

- Try a different mic input (remember to change on the console too).
- Replace the mic cable.
- Try a different mic.

Most of the time it's user error, something simple like not fully plugging a microphone cable in, plugging into the wrong mic patchbay input, missing something on the console I/O module, etc. If you do find a broken cable or microphone, label and set it aside so you can fix it later. If the mic itself is not passing signal after you've tried at least two different cables and verified the console setup, you can look for something on the mic. Make sure the XLR pins are tight in the casing. Sometimes this connector pulls loose, possibly pulling the wires inside. The other common problem results from the mic capsule getting struck by a drumstick or overeager vocalist. Contact your studio manager or service technician.

SIX

EDITING & MASTERING

Once you've bounced your mix to a stereo WAV file from the DAW, open it in an audio editor such as Wavelab or Audacity. These programs are different from a DAW in that any processing or editing is permanent to the file once it's rendered. Here is where you will do final trimming of the waveform and apply any mastering processing that might be needed.

Editing

Editing refers to a lot of different things, some performed in the DAW before bouncing the mix and others in the stereo editor. Here are the more common editing tasks that engineers perform on a bounced mix file.

Song count-offs

During mixdown the engineer won't bother with various noises before and after the song, such as the introductory count-off leading into the song. Once a satisfactory mix is complete, these intros and endings can be easily removed. Just select the extra space at the beginning and delete, then create a very short fade to smooth the start of the clip. Do the same at the end, making sure you don't lose any low-level sounds from a song fade-out.

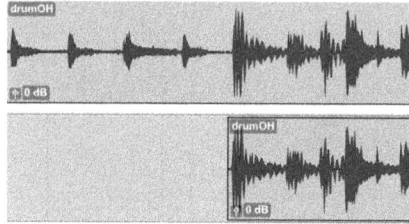

Album sequencing

Once final mixes are completed for all the songs on the album, they must be sequenced into the proper order for the record. Songs are mixed as convenient, whether in the order they were recorded on the multitrack or grouped according to similar mixdown settings. In the old days when albums were sold on cassette or vinyl, we also had to include the break between side one and side two.

There are usually three or four seconds of silence inserted between songs. The audio editor should have a CD setup mode for this, where you can set a certain amount of silence between tracks and arrange the order of songs.

Comping a final mix

Sometimes several mixes are done for a single song. On a console, each mix turns out differently due to various nuances in how the engineers adjust levels, effects, and various other settings. Instead of spending forever trying to get a perfect mix in one shot, a composite mix is built by piecing together the best sections from the various takes. Naturally, significant differences between two mixes may make two consecutive sections unworkable when joined together. For example, if the second mix is much softer than the first at a certain point, then splicing sections of both together may have abrupt level changes that are unmusical. However, if all mixes are relatively consistent this method should work fine.

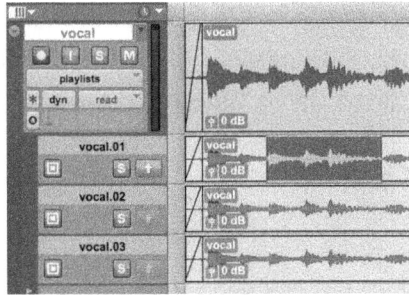

Mixing on a DAW pretty much eliminates this since the mix can be continuously fine-tuned with automation, ensuring consistency during each pass.

Analog tape editing

Not many engineers edit tape anymore, but we'll include a brief overview in case you happen to run into one of these machines someday. Besides, it's kinda fun. On analog tape the unwanted segments between songs are physically cut out and replaced with special, non-recordable tape called leader tape. Leader tape provides a totally silent playback, as opposed to blank recording tape, which outputs tape hiss. You can buy leader tape that has timing marks printed on it, spaced in 7 1/2" increments. Professional and semi-pro tape machines run at speeds either 7.5, 15 or 30 ips (inches per second), which means that the machine is running fifteen inches of tape across the heads each second at 15ips. This makes it easy to insert an exact duration of silence between songs— merely do the math depending on which speed your machine is operating. If you have a 15ips machine and want three seconds of silence, insert six sections of leader tape between songs. Leader tape is also spliced in at the very beginning and end of the reel to facilitate loading the tape onto the machine. Since test tones are usually included on analog master tapes, several seconds of leader should separate the tones from the project material. Before editing software came along, we would arrange the song order for an album by cutting individual song mixes and sequencing them onto a master reel—physically splicing them together to create a master tape. No big deal as long as you remembered which song was on each section of tape. It's not hard once you get the hang of it, but then there was an intern once who we found with bleeding fingers, arm, and even a sliced ear. Be careful with those razor blades.

Labeling & documentation

Always keep an accurate record of what's on your media. This includes song titles, album information, song takes that are keepers, technical details for tapes such as flux level and noise reduction, and so on. Keeping good notes is essential, not just good practice; otherwise you have no idea what's on those hundreds of Pro Tools files and track regions. Probably the number one complaint from mix engineers who receive session files from other studios is poor organization—having to sift through huge numbers of files, audio clips, and so on takes forever and is ripe for miscommunication as to what the original engineers and producers intended for the final mix. They can't read your mind and understand what you intended, so keep it clearly organized so they know what's going on.

In the DAW session, name each audio track before you begin tracking. As you then record multiple takes and edit the track, the DAW will automatically label resulting clips based on the track name. This will save your life over time as you endlessly scroll through the clip list window desperately searching for the third kazoo solo take.

Set markers at major sections of the song, such as verse 1, chorus 2, and bridge. This makes it easy to find any particular part of a song. For tape machines, make notes on the track sheet since data cannot be stored on tape.

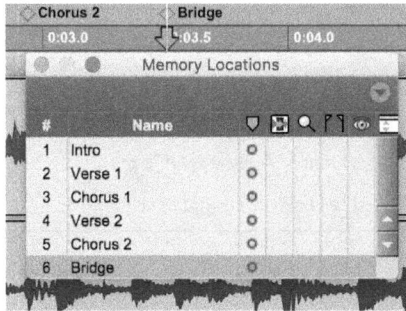

There is a comment box for every track in Pro Tools. Use this to make notes as to mic used, suggestions for mixing, whatever.

The final mix files or 2-track tapes are referred to as *masters* and must be clearly labeled with several items of technical information, depending on the format type of the media. The mastering engineer needs to know certain information about your master so they can set it up for duplication. This helps you as well, because appropriately labeled recordings help you when you dig them out in the future and have no idea what's there. For tapes, stick the label on the outside of the box. For data files include a text file in the folder with any relevant information.

Items recommended for the master label:

- Project title
- Inventory/work order # for cataloging purposes
- Client name
- Date of project
- Engineer(s)
- Studio
- Program material with timing information
- Tape speed in IPS (analog only)
- Recorded flux level (analog only)
- Test tone sequence (analog only)
- Tails or heads out on the reel (analog only)
- Noise reduction? Type? (analog only)

```
┌────────────────────────────────────────────────────────┐
│  Client:                    Project #:                  │
│  Engineer:                  Date:                        │
│                                                          │
│  Tape speed: □30 ips   □15 ips   □ 7 1/2 ips            │
│                                                          │
│  Tracks:      □ Mono    □ Stereo  □ 4 tr                │
│                                                          │
│  Wound:       □ Tails out □ Heads out     NR □          │
│  ──────────────────────────────────────────────         │
│                                                          │
│  Track list                              Time start      │
│                                                          │
│                                                          │
│                                                          │
│                                                          │
│                                                          │
│                                                          │
│                    Hear Real Good Studios                │
│                    101 N Audible Ave                     │
│                    Deafness, ZM 90909                    │
│                     909-555-9090                         │
└────────────────────────────────────────────────────────┘
```

Safety copy & backups

Masters are rarely physically shipped these days, but if you have to send a master somewhere make sure it's not your only copy. So, when the original fails to arrive at the plant, or when your cat's claws scratch the disc, or when the delivery boy drops it down the sewer, you always have a backup and haven't lost your reason to live.

File transfers via Dropbox or other similar services are the most common method for sending masters. Even so, always make sure your data is backed up multiple ways. Have a second copy of your media drive on hand by setting up an automatic backup application that runs every night. Use a cloud-based backup service for remote safeties in case your studio burns down. All of these methods are cheap, so there's no excuse.

Mastering

The final step in the process is to tweak master files so they sound as good as possible and are ready for distribution. Mastering a project involves watching for level issues, such as hot spots that may overload (distort), tonal problems (such as too much bass or treble), stereo imaging and phase issues, and so on. They can also get creative and enhance the audio in lots of ways. A good mastering engineer will save your mix and make it shine, making it worth the money and effort to hire one. If you're serious about your product, don't even think about throwing a cheap peak limiter on your mix bus like you saw on YouTube.

The main components of the mastering process include various applications of compression, EQ, and other processing. These usually entail very subtle, carefully chosen adjustments, and knowing which steps to take requires long experience and highly developed ears. The tools used are also different in many cases; highly precise EQs are used for surgical fixing, EQs with very broad bandwidths are used for overall shaping, and mastering bus compressors are designed to tighten and glue a mix together.

Industry trends can dictate objectives when mastering a project. The most notable change over the past couple of decades is the desire to make final masters as loud as possible. By applying peak-limiters that squash the audio, we can virtually eliminate all dynamic range from a file so it sounds consistently loud. Many engineers and artists decry this practice as it destroys the musicality of the song, but record companies want their product to sound louder than the competition. Here's a before and after example of a mix that was processed with a peak limiter. Note how the waveform loses much of the ups and downs, looking more consistent from left to right and remaining close to the top and bottom edges of the window (close to these edges means higher signal level). These ups and downs represent level changes, which generally translate into perceived volume differences. So, the processed example will not vary much in apparent volume, but rather will sound loud all the way through the song.

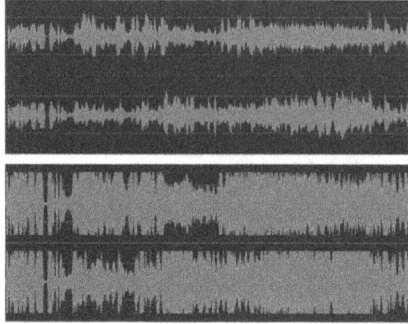

There's no way you can learn how to master audio from this book—or any book. You can learn everything there is to know *about* mastering from Bob Katz's terrific *Mastering Audio*. But to get good at the process you need to set up an apprenticeship with a good mastering engineer and let them show you the ropes. The main thing is learning how to *listen*, which takes guidance and practice. For now, let me give you two approaches for a DYI file prep that'll at least get it in the ballpark.

Normalize with limiting

If you play your bounced mix file in iTunes or Media Player you'll notice it's not nearly as loud as everything else in your library. Part of the mastering process is adjusting levels in a variety of ways to bring the overall volume up to what people are used to hearing. This is what we'll concentrate on here, so first open your bounced mix in an audio editor such as Wavelab or Audacity. Find the menu item for *normalize* and set it for -1.5 maximum peak. It'll then examine the entire waveform searching for the highest level peak, then subtract that from your ceiling of -1.5. If there's a difference, it'll then increase gain on the entire file equally for that amount. So if your file peaks at -4, that leaves 2.5dB of headroom before reaching your max limit of -1.5. The entire file will be increased by 2.5dB, and you'll see the waveform get larger when it's done. This does not change relative dynamics in the file, but merely cranks everything up the same amount. It sounds exactly the same, only a bit louder depending on how much headroom you had to begin with.

Now insert a peak limiter such as Waves L1 or whatever may come with your software. You'll see threshold and ceiling controls. Set the ceiling for -1.5dB true peak, which means there will be no level above this point when it's done. This is important for providing some room for the mp3 conversion process, preventing intersample peaks that distort, and reducing the chance for cheap converters to mess it up when playing back. Play your file and slowly lower the threshold control. It'll get louder, but if you keep going it'll start sounding crunchy. Find a balance between loudness and preserving a clean, musical sound. Render this to a new 24-bit WAV file, but don't replace your original bounced file. Save everything along the way in case you need to revisit any step in the process. Once this is done, use the new file for creating an mp3, AAC, or whatever you need.

Normalize with LUFS

Audio engineers have religiously worried about signal levels ever since recording was invented. The primary issue has always been preventing distortion, which happens when a system or recording medium is over-loaded, while also ensuring sounds are above the system's noise floor. In the days of analog tape, for instance, if you recorded too high it would saturate the magnetic particles on the tape and run out of room, so to speak, for accurately storing and reproducing the original waveform. Record too low and tape hiss would be too loud. Record too high on a digital recorder and it runs out of bits to encode the data, resulting in hash. Engineers still tend to think in terms of peak level, referencing their recordings to digital zero (0dBFS). A peak limiter simply squashes everything upward from the bottom while also controlling the top end so that everything falls into a narrower dynamic range.

The more intelligent approach to all this revolves around how people actually hear and perceive sound. It's not a straightforward, linear expe-rience—our perception of how loud a sound is does not match the phys-ical levels in the audio. And so our digital meters really aren't telling us anything useful about how it sounds to us. Human perception of sound, particularly volume, varies based on frequency. We hear certain frequencies much easier than others, which the Fletcher Munson curves reveal. Listen to a 1k sine wave at a certain level, then play a 100Hz tone at the same level. The lower frequency will be far quieter. Our hearing system isn't flawed; it's designed to optimize hearing speech and other sounds around us. It wasn't really intended to maximize the sub-harmonics blasting out of the dance club.

Loudness meters are designed to analyze and measure audio based on our *perception*. This gives us a much more useful indication of how loud something is, and we can therefore process a file that better matches what listeners actually need. The unit of measurement is the Loudness Unit, or LU, and we label an audio file as being *n*LUFS (full scale). 0LUFS is maximum, similar to 0dBFS, but that's where the similarity ends. Loudness standards have been developed that use this as a guide for ensuring a more consistent listening experience. The goal is for each program to sound about the same as the next. This doesn't mean all dynamics are lost, just that the overall perceived volume is fairly close. For example, public radio established a standard of -24LUFS with a maximum at -3dBFS (-2dBTP for true peak). The unofficial standard for podcasting is -16LUFS for stereo files. This is quite a bit higher than public radio, the reason being that podcasts are typically listened to while jogging, riding the subway, driving a car, and all sorts of other less-than-optimal listening situations. This requires the program material to have less dynamic range, but since podcasts are primarily voice with minimal music, it works quite well.

Music streaming services such as Apple Music and Spotify use loudness control to provide a consistent volume level for their catalogs. Spotify plays back everything at -14LUFS. This doesn't mean that everything sounds equally flat and compressed. Far from it, it actually allows for lots more dynamic range than we're used to. Remember, the measurement is all about perceived loudness, not actual peak levels. Before, we would squash and limit everything to get it as loud and close to 0dBFS as we could, but then streaming services would simply lower that down to their desired level. This doesn't return the original dynamics and musicality to the file; it's still squashed and limited. Master your file with a target of -14LUFS (-16 for Apple Music) and it'll sound far better with more of the musical transients and dynamics preserved.

So, how do we go about this? You'll need a *loudness meter*, which is not the same as all the other standard meters found in DAWs and editors. These meters provide detailed readings including long term (the overall loudness), short term, variation (range), and true peak indication. A classical recording will have lots more variation than a voice-dominated podcast show. Once you get a reading you can go back and apply a gain change to adjust the overall level up or down as needed. For example, your file reads at -27LUFS and it's destined for public radio. Raise the gain 3dB and you should be pretty close. The issue there is that you still need a peak limiter to prevent overshooting -3dBFS. So it's a back and forth process to get it right.

iZotope's loudness control processor not only measures your file, it'll do all the adjustments to make it conform to whatever outcome you need. Again, this is different from a peak limiter that only looks at instantaneous levels to process and control. A loudness processor is looking at a longer period of time and making calculations as to what's needed to conform to a particular loudness level, within tolerances. iZotope explains that their processor rarely compresses dynamics, and when it does it's always in context of this bigger perception picture. The result is a file that matches the desired perceived loudness, but retains most, if not all, its original dynamics. I've turned away from peak limiters for the most part and rely on this tool for all my music and podcast production. The final file sounds much better and it's plenty loud in the car, matching other program material that was produced to the standard.

Take a look at the image below. It's the same file as shown earlier that got flattened by a peak limiter, but in this case there is no limiting. The loudness control plugin makes it sound just as loud as the limited version, but without the crunching and loss of dynamics. This is a huge step forward for voice and music media distribution.

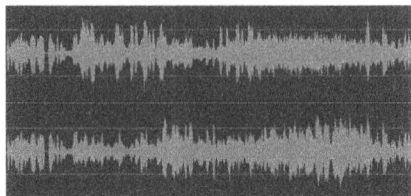

File delivery formats

Always bounce your mixes and render final stereo files in the WAV format at 24-bit. Maintain high quality as far through the chain as possible before converting to mp3 and other lossy formats. Once you make an mp3, you cannot recreate the original quality of the WAV it came from. Save copies of each format as you go in case you need to go back and redo something.

There have been a variety of file formats developed over the years. The biggest distinction are those that preserve all the content and those that eliminate information so as to reduce file size. WAV and AIFF are two common full-quality audio formats. Whatever was captured in the recording is what's stored in the file.

Most everything else attempts to reduce file size by compressing the data file. This is different from audio compression; it refers to looking at the data in a digital audio file and eliminating redundancies and non-critical information. Compression formats that get rid of redundancies, but preserve all the music, are called *lossless*. *Lossy* algorithms are more invasive and affect the music to some extent, trying to figure what might not be missed if it went away. This concept first showed up in the 1990s with Sony's Minidisc and Philip's DCC (digital compact cassette). Perceptual coding attempts to figure out what music content is covered up by other stuff and therefore isn't crucial to the overall sound. A variety of similar approaches have been developed over the years, the most well-known being mp3 and AAC.

These are both lossy formats and offer different levels of compression for you to balance file size with sound quality. Most music is listened to at 256kbits/sec, while a simple voice podcast would be acceptable at 96.

Of course the primary reasons to worry with file size are storage limitations on a listener's device and bandwidth/streaming constraints. Over time this won't be an issue, but music streaming services have to balance quality with how much data people can stream over their mobile data plans.

Which format do you use for your master? Your client or streaming service will specify what they need. If it's a reference so people get an idea of what it's going to be, go with an mp3 or something small. For CD production use the high quality WAV master. Online services such as SoundCloud and so on will have their own guidelines. Podcasts should be either mp3 or AAC (the format Apple uses), though mp3 is the most universal format that everyone can play. So it depends on the situation. Make sure to listen to various settings before you send anything off, so create a few different versions of mp3 and AAC at different compression settings. There are also software tools that will preview your file as it would sound in various formats as well as streaming services, which is very helpful.

The goal is to provide as rich a musical experience as possible within the confines of the delivery medium. Advances in bandwidth for downloads and streaming, better approaches to listening consistency as provided by loudness standards, and improvements in the tools we use for production and playback will continue to move the needle forward for engineers and consumers.

SEVEN

SESSION OPERATIONS

The facilities, personnel, and process required for music production can vary significantly. Much of today's recording is done in somebody's spare bedroom or a rented office. This doesn't necessarily mean poor quality, as many professionals have very nice setups and work independently. Albums can be produced by one person, but are often best completed with the input and direction of competent producers, engineers, and session musicians. This, of course, is a budget consideration since it's expensive to hire professionals.

For music production, a lot of information needs to be figured out before you begin the first recording session. You and your client must meet to discuss various aspects of the project, including the purpose of the project, goals they have in mind, and how they want to achieve them. All parties involved must be in agreement as to what they each expect from this project. Otherwise, if you deliver a product they disagree with, everybody goes away unhappy and they're unlikely to refer other clients to you.

Session planning for the client

Will this be a demo, or is it a full-blown album for distribution and sale?

What exactly do you want at the end? One song or ten songs? Radio play or just copies to sell at gigs? This determines the methodology, quantity of songs, and final delivery medium.

How are you going to make this work? What type of studio do you need? Will you be hiring musicians, an arranger, and a producer? Work out your arrangements before the session; determine what musical parts are needed and which are optional. Make music in the studio, not major musical decisions. Schedule similar songs to record together so as to minimize studio setup and changeover time. What's your budget? Poor planning results in over-budget projects.

Session planning for the engineer

Get familiar with the project. What type of music is it? Instrumentation? Are you familiar with that particular style of music? Can your studio handle the project? One of the few non-label projects I worked on involved a band who visited the studio to see what kind of setup we had. They brought several CDs, listened through the control room monitors, and asked lots of questions—all to determine whether our facility and staff were appropriate for what they wanted to do. I was impressed, and we had a great time working with them. On the other hand, I watched an engineer attempt to record a style totally out of his wheelhouse. It didn't go well.

Make sure you and the client are in agreement with what the final outcome should be; this is a common cause for conflict and budget over-runs.

Plan how to best accommodate their goals. This includes studio setup, mic choice, tracking schedule, possible equipment rentals, and any supplies needed. Do this well before the session dates. Depending upon the studio, some of this pre-production planning and communication will fall upon the studio manager and/or producer. If so, then a work order should be filled out with the necessary information for the engineer and assistants to prepare for the sessions.

Session politics & roles

Producer

It rarely goes well when everybody claims equal input, such as the members of a band. Somebody has to be in charge, and that's the producer's job. This individual must know the particular style of music and should have worked with the group to get acquainted with what they're about. The producer has overall authority for the project and is responsible for:

- Overall production concept
- Business management
- Musical direction
- Artistic decisions

We once did a rare non-label album project that featured a band with seven members—and each one had different opinions. This is normal, but there was no one person designated to make final decisions and so everybody tried to influence the outcome. We spent an inordinate amount of time and money replacing basic tracks and solos, often when existing tracks were more than good enough. I spent an entire night (a solid twelve hours) overdubbing a guitar part throughout one song. There must be at least eighty-seven punches in that five minute track, simply because the player kept changing his mind as we went along. A producer would have stopped this long before dawn, realizing that what we had early on was sufficient, especially considering that the musician was a fabulous player—no need to redo anything. The producer is ultimately responsible to the record company who is fronting the money for the album. The task is to complete projects satisfactorily, on time, and under budget. After all, it's the artist's money that's ultimately being spent.

Many prospective students tell me they want to become record producers someday. However, there is no official certification to be a producer. These individuals learn their craft over time, many while serving as engineers on other projects, then branching out to do a few projects on their own. Gifted producers seem to know just what to do to get the most from their artists and musicians. I was privileged to work on a large number of albums with a man who had that magic touch. Everybody wanted to work with him, and it was easy to see why. Once we had a vocal trio who had just hired a new vocalist. She was nervous and could not quite make it through one particular part of a song. They kept up the pressure and she simply broke down. I got ready to turn everything off and go home, but this producer told me to mute the mic and give them ten minutes. I have no idea what he said to her (I was too honest to listen in), but when he came back we ran tape and she nailed it perfectly in one take.

If you are interested in producing, spend time listening closely to how producers put together the records you enjoy. Even better, listen to lots of different styles—it's rare you get to work in only one particular genre. Producers get paid by the job and find work with bands and companies who like their style and see a successful track record. Good producers not only get paid a flat fee for the project, but can also negotiate points, or a percentage, of the royalty income from album sales.

Chief engineer

This is the primary engineer responsible for listening to the producer and capturing their ideas in the recording. Your job is to always be ahead of the producer. Listen to the music, know where the sections are (verse, chorus, bridge), listen for wrong notes, chords, and intonation problems. Read the producer's mind and know what they'll be doing next. Depending upon the situation you might offer creative, musical input. In smaller projects and studios you might be functioning as both—which absolutely requires musical skills and a good ear.

Chief engineers are sometimes employed by the studio, but often major projects are handled by independent engineers who are hired per job. They therefore work in lots of different facilities, which requires getting used to the nuances between different sounding rooms with different equipment. Spend some time reading album credits and note the names that keep appearing on various projects.

Assistant engineer

These are the silent runners that keep everything moving. They handle studio setup before the session, tear down afterwards, keep notes and track sheets, run the recorders, adjust mic stands, make coffee, and go get pizza. Your job is to always be ahead of the producer and engineer and not get in the way. The hours are long, the demands are high, and it takes a lot of initiative and positive attitude to impress anybody (i.e. keep your job and move up the food chain).

Being an assistant means different things depending on the type of facility. Smaller companies usually provide more opportunity for new engineers to get their feet wet early and do some engineering. Larger, high-end studios may make newcomers wait months before they touch a console, taking the time to observe the recruit and see how determined and motivated they are. If you are willing to stay in for the long haul there are opportunities, but it can be a long, grueling process. In the meantime, keep your mouth shut, ask questions when appropriate, and learn from the more experienced engineers and producers. We had a couple interns over time who didn't understand this concept—they arrogantly thought they should offer their opinions and have never worked in the big leagues since.

Always remember you're there to make music. It's crucial that the artists feel absolutely comfortable when they play. Any uncertainty, frustration, or lack of emotion will be obvious on the record. The mark of a top-notch producer/engineer is knowing how to pull the best out of people no matter what their abilities are. 80% of the music industry is communication skills. Knowing how to translate non-musical, non-engineering terms into your lexicon and vice-versa goes a long way to keeping everybody happy and on track. It's pretty simple really—be nice to the folks and stop trying to prove something.

Session documentation

Any professional business is run on documentation. Sometimes it tends to get out of hand, but you've got to get into the habit of organizing your studio and the projects you work on. Students always ask "But why? It's so annoying and time consuming." It's simple—if you don't track your expenses and time in the studio, how can you bill the client? If you don't keep careful inventory of project media, how do you know which reel or drive is okay to record over? What happens after you've been in business for ten years and have hundreds of files to sift through when the long-time client wants to remix that classic hit from the nineties?

You get the idea. Let's show you a few sample documents that you'll use everyday.

- Work orders
- Invoices
- Instrument setup / mic input list
- Track sheets
- Take sheets
- Media labeling
- Tape logs
- Time logs
- Maintenance reports

Work orders

Your studio is a business, and part of that is making sure you know what's completed, what needs to be done, and when to do it. Here's an example that can be used for album projects that allows all individuals involved to ensure everything is planned for and completed.

Work order

Client: Engineer: Project #:
Contact phone:
Contact address:

Pre-production	Date Completed	Initials
❏ Session booked		
❏ Producer booked		
❏ Musicians booked		
❏ Budget agreed		
❏ Deposit received $		

Recording	Date Completed	Initials
❏ Microphone rentals needed:		
❏ Equipment rentals needed:		
❏ Tracks complete		
❏ Monitor mix to producer		
❏ Monitor mix to artist		
❏ Mix complete		
❏ Quality assurance		
❏ Final invoice to label		
❏ Balance of invoice paid		
❏ Master delivered to label		

	Date
Signatures	
Label representative:	
Studio representative:	

Hear Real Good Studios
101 N Audible Ave
Deaf Valley, NC, 90909
909-555-9090

Mic input chart

Engineers need to keep detailed, accurate notes as they set up for sessions. Write down everything you plug in, including mic inputs, outboard gear patches, etc. A mic input sheet is a simple chart indicating which microphone is plugged into which mic input on the patch panel in the studio. List your mic inputs, which mics are connected, and what instruments or locations they're being used for. Make sure you update this as the session progresses, such as for overdubs or anytime you change mics.

So, why is this so helpful? Once the session begins things can get hectic very quickly. At some point you might have a mic signal disappear or experience some unexpected change in your signal flow. With everybody staring at you waiting for you to fix it, it's your job to know exactly what's going on. Accurate notes will enable you to go directly to the problem to see what's wrong.

Often these charts are provided by the chief session engineer ahead of time so the assistant can have everything ready to go before everyone else shows up.

Track sheet

A track sheet records everything you need to know about each song: which instruments are on which tracks, cue points in the song, general notes from the engineer or producer, project title and studio info, and technical information about the recording itself. This sheet remains with the multitrack masters where anybody can later easily find what's on the tape. For DAW sessions, all of this should be notated within the session file itself.

Hear Real Good Studios	
101 Analog Way Deaf Valley, NC 99999	Client: Producer: Engineer:

Date:	Speed:
Studio:	Level:
NR:	Tones:

Title:				Time:	Cut #		Reel:
1	2	3	4	5	6	7	8
9	10	11	12	13	14	15	16
17	18	19	20	21	22	23	24

Cues:	Notes:

Take sheet

Take sheets are used during the basic tracking of a song. Each time the musicians begin a song, a notation is made as to what counter time each take was begun, whether or not they completed the take, and whether it was a keeper. Do the same in a DAW, where multiple takes are completed in the same session file. Drop a marker (*enter* key) and name it quickly (1 partial, 2 good).

Hear Real Good Studios			Take Sheet
101 Analog Way			
Deaf Valley, NC 99999			

Work Order # _00/0682_   Producer _Frederickso_

Album/Project _LA Jazz_   Engineer _H ll_

Reel/Tape # _00 343_   Studio _A_

Date _7/29/98_

Take #	Time	Code	Remarks
1	:21	CT	OK
2	4:13	FS	—
3	4:29	FS	—
4	5:10	IT	
5	2:33	CT	Keeper – V2 from T1?

Codes: CT = Complete Take   IC = Incomplete Take   FS = False Start

Labels

All tape boxes need a label on the outside so you can easily find something on the shelf. Here's an example:

Client: Project #:
Engineer: Date:

Tape speed: ☐ 30 ips   ☐ 15 ips   ☐ 7 1/2 ips

Tracks: ☐ Mono   ☐ Stereo   ☐ 4 tr

Wound: ☐ Tails out ☐ Heads out   NR ☐

Track list   Time start

Hear Real Good Studios
101 N Audible Ave
Deafness, ZM 90909
909-555-9090

The bottom line is that taking time to be organized and communicating with your client will help ensure a successful project. It may not make the young band sing in tune, but at least everybody will know what's going on and keep sessions running smoothly.

EIGHT

PODCASTING

Podcast recording focuses almost exclusively on voice, so we'll concentrate on that. Unlike recording on site for film and tv, you can usually control your environment by recording in a treated room. That doesn't mean it has to be a million dollar recording studio, but you can at least avoid what too many podcasters do when they set up their laptop in the kitchen or bedroom and start talking at the built-in mic. It's easy to hear the difference between this setup and someone who knows a bit about mic selection and placement, room acoustics, and decent processing, so let's step through each of these. If you want more technical information on microphones, room acoustics, etc, jump over to those chapters.

Microphone selection

Podcasters typically use a large diaphragm mic. Condensers in particular provide clear articulation, sensitive pickup, and a full low frequency response; they capture sound really well, even from a distance. This is both good and bad—you can move around a bit and not lose yourself, but they pick up more room echoes and noise, including the person sitting across the table from you. Some people prefer dynamics for this very reason, and there are several models that are excellent for speech recording. The trick is to stick close to the mic; don't move around or turn your head, or the sound will noticeably change and diminish. But their real benefit is not capturing as much of the room sound around you, resulting in a cleaner track to process later.

There are loads of mics marketed as "podcasting mics", which means, well, not much of anything except a USB connector. Some feature a built-in headphone jack and perhaps a mute switch. Headphones are necessary so you can hear other participants as well as how your voice is coming across. The switch is actually pretty handy for when you get choked up on your twinkies, but it can be added separately. Don't ever use the laptop mic or the cheap headset that came with your Dragon Dictation software; the quality for both is awful, and the laptop mic is too far away to avoid room echoes and noise.

For a group of people talking around a table, the best scenario is for each person to have their own mic. But if that's not possible, use a single mic that picks up all around—not just in front of it. This is called an omnidirectional polar pattern, and some manufacturers emphasize this function by shaping it like a ball (totally unnecessary, but it looks cute). Some condenser microphones feature switchable patterns, so they can be set to omni for a situation like this or cardioid (uni-directional) for standard, in-your-face placement.

A basic $60 USB podcasting mic will do fine to get you started—there's no need to spend $800 or more on a studio recording mic. As you develop your skills, though, you'll begin to hear the difference and will probably want to step up to the next level. Price alone doesn't always dictate good or bad, so search around and find out what professional podcasters are using.

A decent mic stand is important. Cheaper mics come with a tiny stand that places the mic too far away. Look for a guitar-amp stand or something similar, and if you're serious about all this, find an adjustable boom arm that mounts to the desk. These are great because the arm keeps things floating out of the way, placement can be precisely adjusted, and it helps isolate vibrations such as when you bump the table with your knee. It can also be pushed aside when not needed.

USB or XLR?

USB microphones are popular because they connect directly to the computer with no extra hardware needed. I have one that cost about $250 and sounds pretty good. Keep in mind there's certainly a difference between this and a $60 model, so invest the extra money if you can.

The other route is a standard XLR mic, so-called because of the 3-pin audio connector it uses to interface with professional audio equipment. This is an analog signal, so the mic is plugged into either a mixing console or audio interface and must be converted to digital for recording and processing in software. These devices provide a *microphone preamplifier*, which is required to take the very low-level signal coming from the mic and increase it to match what the recorder and other equipment need to operate. Phantom power is also supplied for powering condenser mics; make sure the switch is on unless you're using a dynamic. The other crucial function is the actual conversion from analog to digital, which directly influences the sound you get. Along with the choice of mic, placement technique, and room acoustics, the audio interface is key for sound quality. You get what you pay for, so a more expensive model will have high quality preamps, premium analog-to-digital conversion, plenty of gain, low noise, and should last forever. One mic preamplifier input is required for every XLR mic used; some interfaces come with one, some two, and others might have up to eight or more.

Of course, there are cheap XLR mics that cost $50 and, on the other end, amazing models that cost thousands of dollars. Although I like the USB mic I mentioned earlier, I also have a classic Neumann that blows most anything out of the water...but it runs $3600. And I plug it into a $700 interface. Go for as much as you can afford and don't skimp on the critical components: mic and audio interface.

Don't forget the really expensive mic in your pocket. iPhones can do a pretty decent job if placed in a good spot. It might not be a Neumann, but it'll get the job done. Just make sure you turn it around so the mic is facing the source.

Monitoring

Headphones are essential for monitoring what your voice sounds like as well as hearing anybody else you're recording with. Inexpensive earbuds won't cut it—you need something that will isolate sound around you so the mic won't pick up what's coming through the phones. High quality isolating earphones can work pretty well; headphones must be the type that covers the entire ear. Some mics designed for podcasting provide a headphone jack; otherwise plug them into the audio interface.

Recorders

For recording in the field, portable recorders are very convenient. They're small, record to flash media, have built-in mics, and can often be mounted on a camera tripod. Some models have mic preamps for connecting external mics for better audio quality. You can also take your laptop, audio interface, and studio mic setup. This is more cumbersome, but results in a higher quality recording, all things being equal. Too much stuff? Take your iPad or iPhone along with your good mics; there are recording apps and audio interface options for iOS devices.

All modern recording equipment will record audio at high resolutions, meaning a minimum of 44.1kHz (CD quality) and 24-bits. Read over the digital audio chapter for a technical explanation of what all this means, but always set your equipment at this setting or better. Never record directly to mp3 or AAC, which are highly compressed files; you want the best source material possible that can be processed and adapted as necessary later. It's the same as snapping a photo with a low-quality jpeg setting and wondering why it's so pixelated when printing to 8x10 stock.

Watch the meter and set the incoming mic levels so they're not close to hitting zero (top of the scale). 0dBFS (full scale) is the absolute maximum for a digital recording system; anything beyond that turns into hash as the system runs out of bits to encode the signal. Though you don't want to record too low, which results in more noise, it's not necessary to push the top limit like we did in the old days. Current technology has more than sufficient dynamic range (quiet to loud), so just get it safely in the upper-middle range and you'll be in good shape.

For computer recording lots of podcasters start out with Audacity. It's free, after all, which is a pretty good incentive. Once you get a quality conversion from the microphone through your interface, it's less critical what application you use for recording and processing. Garageband also works fine and is also free (once you buy the Macintosh). Audio interfaces are usually bundled with a DAW (digital audio workstation). Any of these will work, but once you become more sophisticated and have advanced needs for editing and processing you might step up to something like Pro Tools or Logic Pro. Both of these are very powerful and come bundled with all the processing options you need.

The computer recording signal chain goes from microphone to audio interface, which connects to the computer via USB, then back out to the interface for monitoring through headphones.

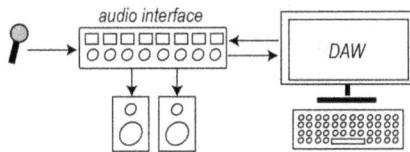

Multi-location recording

Many podcasts are recorded with participants located in different cities (or countries). Skype or telephone interviews are possible using software designed to capture these signals. Browser-based solutions are also available that synchronize everything.

The goal is to have a separate audio track for each person in the conversation so you have control over each one in the mix. Your guests won't need fancy software for any of this; they just need to record their voice. The simplest way is for you to run the headphone jack (which has their side of the conversation) to a recorder or audio interface. It's better if they record on their end, so if they're on a Mac have them use Quicktime, which is capable of making an audio-only recording. Click the small arrow next to the record button to select the microphone as the input device, then press record. The key is using a decent mic rather than their laptop mic. With some guests you won't have a choice, so do the best you can.

One trick for syncing everybody in your DAW during editing is to use a master recording from your Skype session, which includes everything, as a timing reference. Import the individual tracks from your participants and visually match their waveforms. You can also try having everyone clap on a countdown at the beginning, but ultimately you'll adjust here and there as necessary for everything to feel right, especially for longer shows.

Microphone placement & room acoustics

Closer is better for voice recording. Avoid putting the mic a few feet or more away from a person, especially if it's you as the main host. The goal is clear articulation, full frequency response, and minimal interference from the surrounding acoustics. The challenge is that microphones don't behave like people when listening in a room. If someone is talking across the room from you, it's pretty easy to focus in and hear what they're saying. We don't conscientiously register what the room is doing to the sound as it travels across.

Mics, on the other hand, are incredibly stupid—they get what they get with no bias or judgement capability whatsoever. If a source is very close to the mic, it will capture mostly that sound with little of the surrounding room environment. But move the mic just a little farther away and things begin to change.

When a sound is generated in a room it spreads throughout the space until it hits a boundary, such as a wall, ceiling, filing cabinet, or window. It then bounces back until it hits something else. Eventually the energy dies away, depending on how reflective or absorptive the room is. This is the difference between a hard-tiled bathroom and a living room. The comfy couch, carpet, plants, and curtains absorb sound energy and reduce reflections. This results in a drier, more controlled environment, which is ideal for voice recording.

Sitting in a kitchen with a laptop a couple feet away captures lots of room reflections, making the voice sound less clear, more reverberant. The "echoey" room sound will reduce intelligibility and make your show sound unprofessional. Even with a quality mic, if it's two to three feet away you'll get too much room sound. Another issue for distant mic placement is that sound energy attenuates (diminishes) over time and space, so the mic is getting a quieter sound. This has to be amplified in your recording software, increasing noise and the surrounding room sound. Finally, every room has background noises you've tuned out long ago—air conditioning vents, refrigerator hum, cars driving by. Condenser mics get all of this, but less so if the mic is pretty close to the main source. So proximity is key.

Place the mic a few inches from your mouth, perhaps angled just a bit over to one side to avoid direct "pops". What's that? Well, first check to make sure nobody's looking. Now hold your hand in front of your mouth and say words with "p" and "b" consonants. You can feel the puffs of air, especially downward from the mouth. This energy "explodes" across the mic's diaphragm, causing a low-frequency pop in the sound. A pop-filter, usually a round screen made of nylon or metal, is helpful to reduce these pops as well. Mount this just in front of the mic without touching. When talking, stay pretty close to this position; a change of only a couple inches or so will make a difference in the sound, so try not to get too excited and dance around.

An omni mic sitting in the middle of a group will do a decent job picking up everybody as long as they're relatively equidistant from the mic and not too far away. You'll get lots of room sound, so it's not as up-close and personal as a regular mic placement. Sometimes you have no choice, though. It's especially important to treat the room itself so as to reduce reflections.

Fixing your room

Most professional voice recording is performed in an acoustically treated sound booth. This usually means there are absorptive panels on the walls and perhaps on the ceiling as well. The room is also fairly small, so it can't generate lots of room reflections and reverberation. Even if you don't have access to such a facility, there are fairly simple solutions for improving the space you have.

If this is an area dedicated for recording, it makes sense to invest in some acoustic sound panels for the walls. They're not expensive and can be ordered in a variety of fabric colors. It's not necessary or advisable to plaster the walls from one end to the other. Flat, bare wall surfaces are the enemy, so just spread them around the room to minimize large sections of untreated area. Other options are to hang curtains (thicker is better), install artwork that has depth and various dimensions, fill bookshelves with books of various sizes, or even roll a coatrack over with your winter collection.

Avoid sitting close to a wall; not only can these reflections make the voice sound reverberant, they can also destructively affect the quality, or tone, of the voice. What happens is that the sound waves of your voice go directly to the microphone, but also toward any nearby wall. Depending on the angles involved they'll reflect back toward the microphone. This is the same sound, but delayed in time because it takes longer to reach the nearby wall and over to the mic. If you remember anything about sine waves from school, take two copies of an identical signal. Now slide one over a bit in time and add the two together. As the various positive and negative points line up they'll construct a very differently-shaped signal that doesn't look, or sound, like what you started with. This is what happens when a nearby wall, filing cabinet, or other hard surface reflects sound back into the mic. A simple acoustic panel will do the trick, and again, staying closer to the mic will reduce the impact.

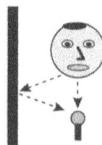

Processing

Once the voice track is complete, you can pretty much follow the steps I outlined in the first chapter of the book. EQ is always required and usually consists of a low-cut filter (set it around 100Hz), some low-mid attenuation to reduce muddiness, and perhaps a bit of high-mid boost for additional presence and clarity. I hear lots of podcasts where the guy phoning it in sounds muffled in comparison to the host sitting in their studio; this gets difficult to understand while driving down the freeway. Try to match tone quality as best you can to make it easier to hear both sides of the conversation. A few dB of compression will help even out the dynamics, meaning less variation in volume as they talk. Review that chapter and you'll be pretty close.

Audio example 68: Processed voice recording (before/after)

Editing is necessary to get rid of dead space before and after the recording, add smooth fades to get in and out of the show, and perhaps deal with unintended starts, stops, and stutters along the way. This can be quickly accomplished in your DAW; again, go look at the first chapter.

Sometimes more extensive work is required depending on the nature of the show, how it was recorded, and so on. Let's say the show consists of an interview between two or three people who are in the same room. Each person has their own mic, and they're all sitting around the table. Go back and listen to one mic solo'd in the DAW, then turn the other mics on. You should hear the voice change a bit and sound somewhat more "roomy". Each mic is picking up everything in the room, including the other voices at the table. So when one person is talking, the other mics are capturing the sound, but from a greater distance, and therefore with more room acoustics. If you decide this is too noticeable and needs to be fixed, or if you just want to clean up tracks when someone's not talking, there are a couple of approaches to try.

Advanced DAW applications like Pro Tools and Logic have a feature called "strip silence". You set a threshold, which means anything quieter than this will be eliminated. What's left should be the actual talking from that person. The trick is that in a conversation where everyone is in the same room there are all kinds of noises made in response to each other, such as "mmm", "uh huh", and short bursts of chuckling or laughter. It's often very difficult to set a silence threshold that distinguishes these from leakage of the other voices in the room, so if strip silence is too aggressive you can try a noise gate.

Gates are processors applied to individual tracks that will reduce volume automatically when a signal falls below a certain level. So, you're talking during the intro, ask your guest a question, then wait for the answer. A gate will open up the channel while you're talking, but as your last word fades away it'll kick in and attenuate the channel so you don't hear the background noise. Unlike strip silence, a gate is variable in how severely it shuts down this noise. It can be set to effectively turn off the channel, like strip silence, but it could also merely reduce the track volume a bit. This might be enough to lower distracting sounds while allowing interjections and such to come through.

Yet another option is to automate the track volume throughout the show, turning the fader up and down as desired as you go along. The idea is to set the volume level up full when the person is talking, down when not, but allow for these low-level interactions that can be important. The most effective and accurate way to achieve this is by manual adjustments. The diagram below shows a two-track interview that has been automated. See the fluctuating line below the track's waveform? That's the volume, and sometimes we set it to pull the fader all the way down, other times just to reduce it a bit so it's not a complete mute. You'll also notice "fades", where the line doesn't just suddenly turn on, but ramps up and down. This smooths the transition, and the key is listening to all tracks together to see how each contributes. Recording automation like this is far more time consuming than using a gate or strip silence, so you have to decide what's worth your time for the desired show quality.

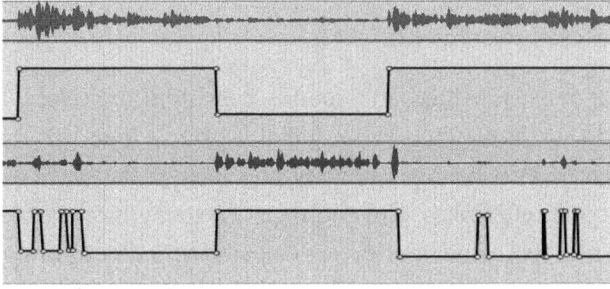

Now, if everyone is in different rooms or locations, this becomes much easier since there's no leakage between mics. Try the strip silence approach or just listen to everything together and see if you need to do anything at all.

Audio example 69: Using automation to reduce mic leakage (before/after)

There are times when people start talking at the same time, especially if everyone's in a different location and can't see each other. You could just delete the less-important outbursts, but if each bit contributes to the show, just isolate each region and slide them apart from each other in time, moving everything else in the show farther along. Steve starts, then fade into Jane's question, then perhaps over to Chris, and the listener has no idea it wasn't planned that way.

Once your tracks are edited and processed, make sure the volume balance between everything sounds natural. Avoid having the host sound loud and close, then have to strain a bit to hear the interviewee. And unlike music production, you almost never want to add reverb to a podcast recording. Doing so muddies the sound, lessens intelligibility, and sounds unprofessional.

Bouncing and mastering

Once your show is edited and sounds consistent and clear, bounce it from the DAW as a stereo WAV file at 24 bits. Open this in an audio editor such as Wavelab or Audacity and look at the waveform. You'll probably see a few spikes here and there, so zoom in on each one, highlight that narrow region, and process a gain reduction to bring it in line with everything surrounding it. It might only need a couple dB or as much as 6. Audition the change to make sure it's not audible, then move on to the next spike. What this does is increase your overall headroom before normalizing the file. Do that next, which finds the highest peak, subtracts that from zero (max), and then raises everything that same amount. Use -1.5 as a ceiling.

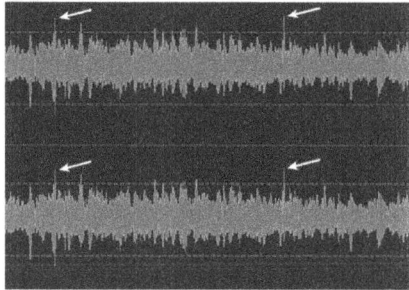

Now you can do a couple of things, following the steps outlined in the mastering chapter. Insert a peak limiter, set the ceiling to -1.5, and bring the threshold down until it sounds louder, but not crunchy. Render this to a new file and use your loudness meter to see how close to -16LUFS it is. A simple gain change in the editor can raise or lower it to match -16, but you have to be careful not to exceed the -1.5 ceiling. This is a back-and-forth process that's time consuming, but can get the job done.

My preferred way, if you can get a loudness control processor such as RX Loudness Control from iZotope, is to avoid the peak limiter altogether. Render your file with the gain changes and normalization we just did, then open in your DAW. RX Loudness Control works as an audiosuite plugin in Pro Tools, so in this case load the new file into a blank Pro Tools session, highlight the entire track and open RX Loudness. Set your target to -16LUFS, make sure the gate is enabled, and process. Done. If you've done a good job mixing the original show session, meaning balancing your different tracks and using volume automation to reduce hot spots and bring up low levels, then the final file will sound consistent. My advice is to spend time mixing in the DAW rather than squashing it to death with a peak limiter, but that takes a lot more time. If you're cranking out multiple shows a week that might not be feasible, but over time you can create your own work flow that balances all of this.

Exporting the final show

When everything is ready, the show has to be exported as a final audio file, ready for uploading to a podcast hosting service. Even though you always want to record and edit uncompressed, high quality WAV files, the finished product must be compressed and in a format that works on listeners' media players. Most podcasts are distributed as an mp3, a highly-compressed format that reduces huge CD-quality files to a fraction of their size. They work quite well for predominantly voice productions like this. The other option to consider is the slightly better AAC (m4a), which was adopted by Apple for their iTunes service. Most players can handle either, though mp3 is guaranteed to work pretty much everywhere.

Set the mp3 encoding rate to at least 96kbit/s. 256 is preferred for music, but this will do fine for voice files. The higher the rate, the larger the file, which translates directly into higher costs for you as many hosting services charge by data uploads. It also makes it more cumbersome for users to download and store. So don't go overboard if you don't need to. Experiment with different settings, load it into your phone, and go jogging to see how each compares. Or in my case go sit on the couch and cycle through them with a good cup of coffee.

A dialog box will appear asking for various metadata tagging information; go ahead and fill that out as it helps the end listener know more about the show, find it in their catalog, and so on. You probably want to enter the title of each episode in the title field, along with the show number and/or date, and use the album slot for the main podcast show title. This sorts everything together in iTunes and other players. Import the show file into iTunes so you can set episode show tags and add an image logo (File > Get Info from within iTunes).

There are software applications (*Forecast* is what I use) specifically designed to encode (export) podcasts while adding helpful metadata such as show notes and chapter markers. Chapters are handy so the listener can jump to, or skip, something in the show. Recurring tasks such as show tags are automatically completed, any markers you set in the DAW file become chapters, and in general it takes care of the entire process. To use one of these, export (bounce) your show from the DAW, but don't set it for mp3. Keep it as a high quality WAV and then bring that into the podcast encoder app.

Posting and hosting

I'm not going to even begin listing and comparing all the various hosting services. Just know that for people to find, subscribe, and download your shows, there's a particular way they have to be made available. The easiest is to buy a monthly subscription with a hosting service; different pricing tiers provide increasing levels of data uploads and other variables. But to test or enjoy your own show, remember it's just an audio file. Copy it to your iTunes library like any other music file and play away.

Hardware and software summary

So what exactly do you need to record a podcast?

- Microphone
- Mic stand or boom
- Pop filter
- Audio interface (if not a USB mic)
- Mute switch (optional, but handy)
- Headphones or decent isolating earphones
- Portable recorder or computer with recording software

- XLR/USB cables

As we mentioned, there are lots of options these days for equipment. Some of it is inexpensive but pretty decent, some of it is awesome and practically unaffordable for mere mortals, and much of it is, well, cheap. My motto is to always buy as high quality as I can afford. It'll last longer, work better in the meantime, and just feels good to use. Splurge on a decent mic and interface, then upgrade your software later. When your show starts generating tons of Patreon dollars you can step up to that classic Neumann or whatever...or just buy a Porsche instead.

NINE

FILM, TV, & VIDEO

Audio for video has come a long way, but still gets second-fiddle status in the real world of budgets and time tables for production. With all the complications associated with recording audio on location and fixing it in the studio, it can be very challenging to produce a final soundtrack that sounds like 1) it was recorded in the space the viewer sees on the screen and 2) that it all happened together at the same time. Remembering fundamentals of room acoustics and microphone technique will go a long way toward making that final mix less painful, so let's take a look at some typical situations and what sound recordists typically have to do in order to help bring a show to life.

On-set recording

Movie and tv sound typically starts on-location by capturing actor dialog. But "on-location" means they're shooting outside on a city street, inside a large soundstage, or up in a helicopter. You've got actors constantly moving around, trying to run/dance/fight/breathe while saying their lines. Acoustics, mechanical noises on set, and planes flying overhead are constant issues. So how is it done?

If the actors are relatively still, a boom mic works well. A large-diaphragm condenser is attached to a long boom and hovers over the group; the operator's job is to keep the mic as close to the actors as possible while staying out of the camera shot. Sometimes this isn't feasible, and so lavalier mics are attached to each person. These are small, clip-on mics like you see on the news, but of course they have to be out of sight. Some models are incredibly tiny and can be hidden inside various pieces of clothing. A wired connection is always preferred, but this restricts an actor's mobility and might be seen in the shot. Wireless is a last-resort due to their unpredictable operation; batteries, radio interference, dropouts, and so on are a constant headache. If nothing else works you can mount a small portable flash recorder and transfer the file later. Sometimes a lav or other condenser mic is placed in a nearby plant or object; this will capture anything close by, but won't have the same intimate sound as a mic directly on the person. It might also pick up reflections from around the object, so listen closely to see how clean the sound is.

Lavalier microphones should be positioned roughly a fist-sized distance from the mouth. Too far away and it sounds more distant and less like you're hearing the person talking to you. Too close and you get pops, low-frequency proximity response boost, and more variance in the tonal quality when the actor turns their head. Even better than lav mics are head mics that mount over one ear and position the mic capsule to the side of the face. These sound fantastic and don't change tone as the actor looks around since the mic is effectively attached to the face. Of course the main issue is that they're visible on that side, so it's necessary to block the shot appropriately.

Lav and head mics are usually taped in a way that minimizes getting pulled out of position while the actor moves around. Wires have to be run in a way that they don't rub against clothing or else they'll generate noise. Wireless transmitter antennas should not be looped and stuffed into a pocket.

Our golden rule for microphones and acoustics always applies—closer is better for clean, intimate sound. The farther a mic is, the more it will capture surrounding room acoustics, echoes, destructive reflections, and other noises. Moving a mic even a few inches farther away can have a dramatic impact on what you get at the recorder. The inverse-square law of physics tells us that every time a sound source is moved twice as far away, the sound level decreases roughly 6dB. This is very noticeable, so consider the change with a mic that's currently two inches from the mouth. Move it down another two inches and you've lost 6dB. Move it down another four and now the mic preamp has to be cranked up 12dB to compensate. That's a lot of gain on the level pot, and that's not the only thing that's changing. We mentioned in the podcasting chapter that mics aren't intelligent like humans (or most of them), but what that means here is that we're pretty good at ignoring extraneous sounds and focusing on the person speaking to us. Microphones can't do this, so from farther away they'll capture everything equally—the person speaking to it, the air conditioner vent, the film camera, echoes in the room, and so on. So close mics are best, but visibility and mobility will always supersede the audio.

Very few film sets are built with well-controlled acoustics, so in addition to room echoes and reverberation another issue will be phasing caused from destructive reflections. As people talk in a room, sound travels around and hits a nearby wall or object. Depending on angles, reflections from this bounce into the microphone, interfering with the direct sound from the person. This changes the tone, making it sound hollow. In a recording studio steps are taken to break up these reflections through a combination of absorbing and diffusing (scattering) soundwaves. On-set options are limited. Try to avoid actors speaking directly toward a wall; angle the entire setup toward a corner or open side of the set. If blankets or other soft materials can be hung on the walls that helps a great deal.

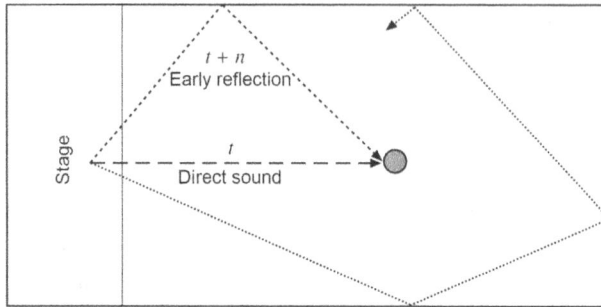

Fixing dialog

No matter how hard you try, most likely a good percentage of dialog recorded on location will have issues that require fixing or replacing. A line of dialog might be muffled from a scarf over the lav, the actor was jumping over the couch and didn't quite get the line cleanly, the actor got punched in the lav instead of the chest, a 747 roars overhead while shooting that Civil War special. The bag of tricks for fixing audio in post-production is quite extensive, but here are a few basics.

Noisy dialog tracks, even with background noises such as the Boeing, can often be cleaned up with software. Tools such as iZotope RX are nearly magical for enabling an engineer to essentially erase something right out of the waveform. A botched word of dialog might be repaired by snatching a similar syllable or word elsewhere and blending it in. But when things are irrecoverable, it's time to bring the actors back into the studio to re-do their lines.

ADR, or automatic dialog replacement, is a tedious process where the actor watches the video of the scene and reads the script over and over until they get it matched just right. This is also called looping dialog. The film director will coach the actor on the delivery, getting the audio to match the visual on screen. It's a much more controlled environment, both acoustically as well as focusing attention on the dialog, so you can usually get a cleaner track. But it's out of context, so it can be difficult for an actor to feel like they're "in the moment".

The challenge for the audio engineer is to then match this clean recording with what's being kept from the original shoot. It all has to sound seamless and in the same space, so level matching, careful EQ, and artificial reverb is used to blend it together. One trick is to record a minute or two of "room tone" on location, meaning after everyone else is gone from the set, record a track of silence. Every space has a certain sound, even when nothing seems to be going on. Listeners can tell the difference when this is mismatched during a scene, so room tone can be used to blend in ADR edits.

Post production

Edit first, mix later. Import the final video cut into the DAW and start putting all the dialog together. Clean up words and match room tone as mentioned above, and avoid the temptation to throw an EQ (or anything) on a track. At this point just get everything to flow well and don't worry about tonal adjustments and so on. There's no context for this until the entire project starts coming together during mixdown. Align all the audio clips and tracks with the scene. This can be done by moving the clip regions around, or they can be located using the *spot* feature, which anchors a clip to a specific frame location. Do the same for sound effects and music—get it synced to the video, smooth over any edits with crossfades, and make any other changes dictated by the final cut of the video.

Mixing for film is done on a dubbing stage, which looks like a movie theater with a large console and other gear in the middle. The idea is to replicate the viewing and listening experience moviegoers will have in an actual theater. Most everything else is usually done in a studio. These types of sessions typically involve hundreds of tracks, so it's not feasible to just start mixing everything at once. Start by mixing dialog, for instance, getting all these tracks blended together, then either assign them to a group or output as a separate *stem*. Stems are sub-mixes of major components for a video project, such as dialog, sound effects, and background music. It's much easier to sub-mix these categories, then bring it all together. At that point if the dialog needs bringing up you don't have to deal with a dozen tracks. Grouping tracks in Pro Tools allows you to control various functions for a number of tracks with a single fader (or whatever). You can also create a VCA track that will control level for any tracks assigned to it. This is a handy way to quickly adjust the level of all dialog, for instance.

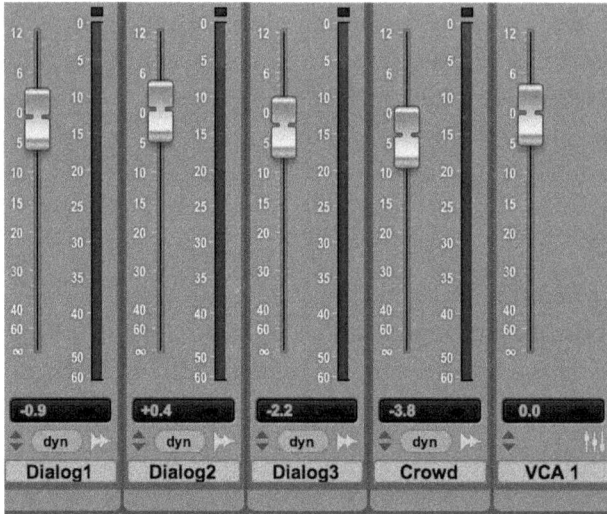

Individual clips can be raised or lowered with the clip gain function in Pro Tools. There's a small fader icon in the lower-left corner of each clip (zoom in a bit if it's not showing). This is far easier and quicker than recording automation, and it's extremely handy for bringing up a bit of low-level dialog or slightly reducing the impact of a cannon.

Color-coding tracks in a DAW is a great visual organizer, so set all dialog tracks to one color and so on. And of course make sure dialog or music tracks aren't scattered all over the place. Keep each type grouped together.

Dialog is always mixed to the center channel when working in surround. This is different than merely leaving the pan pot in the middle for a stereo mix. The idea is that everyone in the theater needs to hear the dialog, and if voices move over toward one side it leaves the folks on the opposite side wondering what's going on. Put dialog in the center, then move music around it with the left and right channels. Sound effects will be wherever; just don't cover up the dialog.

Need extra presence on a dialog track? Boost a bit around 2–4k or so. More warmth? Push up in the 125–250 range, depending on the voice. As we discussed for mixing music, watch muddiness in the low-mids. It all accumulates and creates a muddy-sounding mix. Want to move something farther back into the scene? Roll off the highs a bit, maybe add a touch of appropriate reverb. Add a little low end to bring it forward along with a slight high end boost.

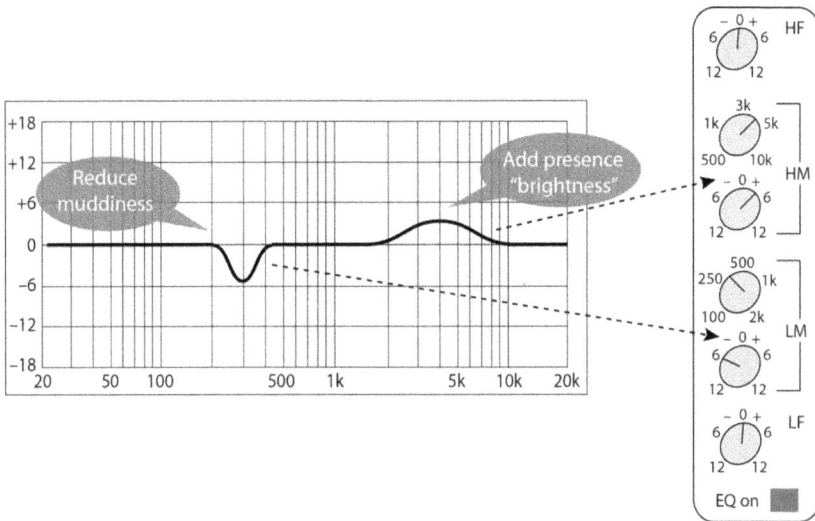

Part of the fun of creating audio for video is coming up with various sounds and effects. Maybe the lead is on a radio headset reporting back to headquarters. You can run a plugin that will distort and destroy the signal to make it sound like a wireless transmission. Or Darth Vader shows up and starts hissing and bellowing at someone. This is entirely different than music production; it's great fun, but takes a lot of time to learn the tricks and experiment.

When you think you're finishing mixing, take a long break or come back the next day. We tend to lose perspective after listening and working for a long time—it will always sound different later. When finished, layback the audio to the video by exporting the session as an AAF file. This format is intended to facilitate sharing audio and video sessions between various software applications. Your Pro Tools session is now ready for importing into Adobe Premiere, Apple's Final Cut Pro, Avid Media Composer, and so on.

All of this really just scratches the surface, but it should give you a good starting point. If this type of work is your thing, check out Jay Rose' *Producing Great Sound for Film and Video*, which has extensive procedures, techniques, and tips for producing film and tv sound.

TEN

RECORDING CONSOLES

A recording console is the heart of a studio—everything converges here, including microphones and the equipment in the control room. Even with DAW-based studios, the concept of a console is maintained in the software. This is where you run mic signals to your multitrack recorder, set up a cue mix, insert processors and add effects, and do the final mix. The key is learning *signal flow*, which is how audio signals are routed during tracking and mixing.

There are countless models of consoles and DAWs, but the fundamentals are the same. As a novice all you see is a mass of buttons and switches; once you get some experience you're beginning to see patterns. This means that you see operations, not simply controls. To run a mic signal to the multitrack your mind automatically filters out everything except for what's needed to route it. Take a look at this graphic:

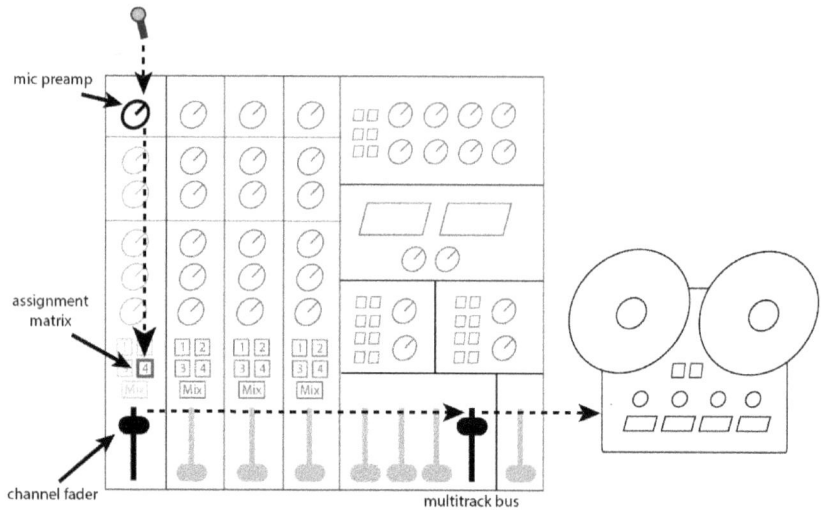

It highlights only those controls needed for a particular operation, in this case routing a mic signal to the multitrack recorder. All other items are dimmed, exactly what happens in your mind as you work at the console. Are you really good at chess? I'm not, so I look at the game board, see two rows of various medieval miniatures, and try to remember the rules for how each type can move. An expert chess player takes a quick glance and sees patterns of strategies, including not just the next move, but several moves down the road. Experts think in terms of operations and patterns, exactly what you will develop over time at the console...but only if you understand how these operations work and practice at it. The value is that it helps you operate more efficiently, be able to solve problems as they arise, and know how to figure out new situations. Consider a scenario where you can't get a mic working. Instead of looking at each row of knobs and buttons, you quickly scan the controls directly related to routing a mic signal to the multitrack. You can instantly tell if you missed a switch somewhere—very quick and efficient. Over time you'll be able to walk into any studio, sit down at their console or DAW, and know what you're looking for without starting from scratch.

Console types & design

Analog consoles are generally classified as large-format or small-format models, the larger boards being the ones you see in major studios. Two of the main differences concerns the number of channels and multitrack busses; you'll see designations such as 24x8x2 indicating channels, MT busses, and the main mix out (L/R). Twenty-four channels allows you to play a 24-track recording back through individual channel modules for mixing. Multitrack busses are the output connections that feed the tracks on your multitrack recorder. So if you're using an 8-track recorder or an audio interface with eight inputs, you would want a mixer with eight separate outputs that feed each individual track input. All consoles also have a stereo (2-channel) output bus which serves as the master mix output. This is connected to your 2-track recorders (where your final mix is recorded), as well as your monitor system in the control room. The large boards typically run anywhere from 24 to 100+ channels, provide 24 MT buses, and take up a lot of space. Smaller consoles have eight to twenty-four channels or so and may only feed four or eight discrete busses.

Large-format consoles provide more controls and flexibility, perhaps dynamics processing on individual channels (compressors and gates), built-in patchbay, and increased options for routing signals. They also usually sound better, but that depends on the manufacturer, design, and of course, price. You can find several small, very high-end consoles that cost nearly as much as a large board, so size doesn't necessarily mean anything.

There are two primary design configurations for recording consoles. The *split console* has channel input, bus output, and track monitoring functions located in separate sections on the board. They are easy to use, but take up more space since everything is spread out. *Inline consoles* integrate input, output, and monitoring on the same channel modules. This design approach saves space, but can be more confusing until you get the hang of it. Most boards are inline designs, but then there are hybrids that feature aspects of both.

Inline signal flow

Inline recording consoles feature two signal paths in each individual channel (also known as an *I/O module* for input and output). One is for incoming microphone signals (called *mic, channel,* or *record* path) and the other is for returning the track from the recorder for monitoring (*monitor, tape,* or *DAW* path). Sometimes this will be the same audio coming in twice: first from the mic to record, then immediately returned from the multitrack for monitoring. Often, however, we'll have two different signals running through a channel. Let's say you record a vocal to track one and the mic is plugged into channel one. Easy enough, but now we want to do an overdub to a different track. Keep the mic in channel one, so you retain all your levels and processing, but simply reroute the mic to a different track using the assignment matrix. Now you have the original track returning on channel one monitor path, but the overdub part coming in channel one record path from the mic, going to track two on the recorder, then returning on channel two monitor path. The diagram below can help visualize this easier.

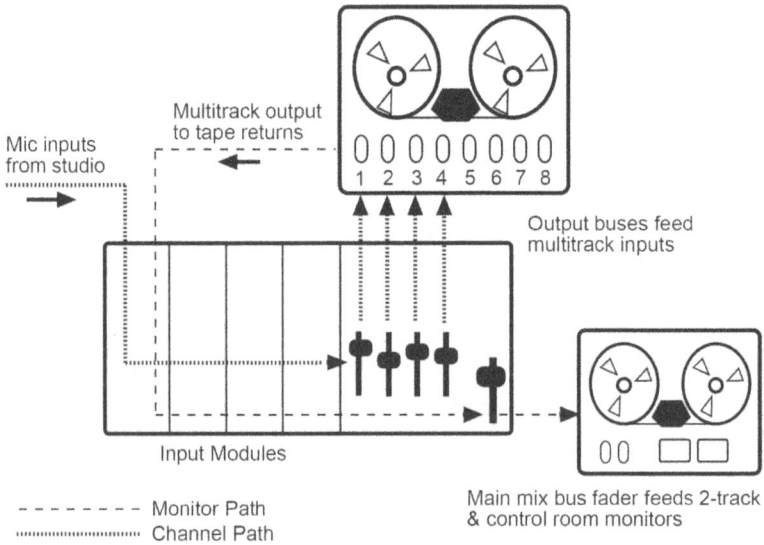

Mic inputs from studio

Multitrack output to tape returns

1 2 3 4 5 6 7 8

Output buses feed multitrack inputs

Input Modules

Main mix bus fader feeds 2-track & control room monitors

- - - - - - - - Monitor Path

.... Channel Path

Many of the various controls on a channel, such as EQ, aux sends, and inserts, can be switched into either of the two signal paths. This means you can put the EQ into the channel path to adjust the audio as it's being recorded. Or you can source the aux sends from the monitor path for generating reverb on tracks from the recorder.

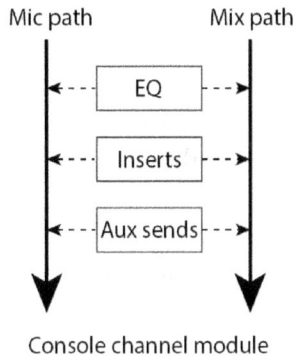

Mic path   Mix path

```
   |<- - -[   EQ    ]- - ->|
   |<- - -[ Inserts ]- - ->|
   |<- - -[Aux sends]- - ->|
   V                       V
```

Console channel module

Console operations & signal flow

Tracking sessions

There are three primary console operations to distinguish during a tracking session. Keep these separate in your mind as you learn your way around, meaning just focus on one operation at a time as you set up the console.

- Route microphones to the multitrack.
- Set your monitor returns from multitrack to the control room monitors.
- Send a cue mix to the tracking room.

Routing microphones

Mics are connected to the channel preamplifiers (mic inputs), which increase signal gain to line level. Once it's in the channel, you have to specify where it's going to be recorded on the multitrack using the assignment matrix. This is a series of numbered buttons that correspond to the tracks. There's usually a channel fader for gain adjustment as well, then finally a multitrack bus level control in the master section that directly feeds the recorder.

Setting up monitor returns

You've got to hear everything you're working on, so audio tracks from the multitrack recorder come back into the console channels, are routed to the main mix bus, and then go out to the main monitor speakers. Note that these signals correspond to the tracks on the recorder, not where the microphones are plugged in. So if you have a mic plugged into channel one and routed to track five, the monitor return will be on channel five. We explained this earlier in the tracking session chapter.

Cue mix

A cue mix is a separate mix of your tracks that's used for the studio musicians to play by. You don't want to send your main control room mix, because it'll keep changing as you and the producer make adjustments, solo tracks, and so on. Since aux sends are essentially independent mixers, they're perfect for this. Turn up the same aux on all your channels, source them from the monitor path, and route the master aux bus to the studio cue system.

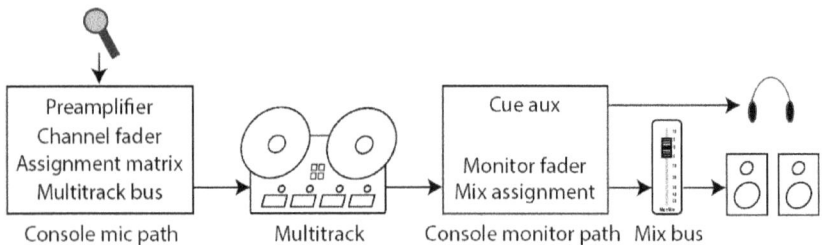

Preamplifier	Cue aux
Channel fader	
Assignment matrix	Monitor fader
Multitrack bus	Mix assignment
Console mic path	Multitrack	Console monitor path	Mix bus

Mixdown

Mixdown is much simpler in terms of signal flow, involving audio tracks from the multitrack recorder returning into the console channels, over to the mix bus, then out to the 2-track recorder and monitor speakers. Along the way you can add various processing such as EQ, compression, and effects via inserts or aux send & returns.

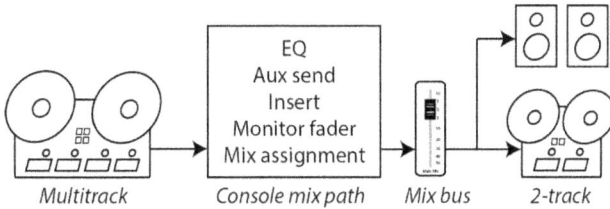

Multitrack   Console mix path   Mix bus   2-track

DAW signal flow

If you're running a DAW, the main difference is not having two signal paths running through each track. What happens instead is the mic signals don't go through your "console" first, but are recorded directly to the system on whatever track it's plugged into on the interface. As you record, the signal shows up on your DAW tracks where you can monitor and set up a cue mix just like on a console. What you see is *post*-recording, so manipulating anything on the DAW track won't affect the recording being made. Of course when mixing, playback simply runs tracks through each channel and over to your mix bus/main interface output.

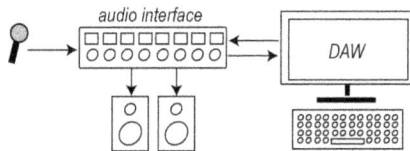

audio interface

DAW

Cue mix on a DAW

It's recommended to set up a cue mix in a DAW just like we do on a console. Instead of routing your main mix to the headphones, give musicians their own custom mix using sends (aux). Route these to two different physical outputs on your interface (it must have more than two for this to work), then plug that into your headphone system.

Console controls and functions

Input channel (I/O module)

Mic input

Every channel has at least two inputs: mic and line. Recording consoles also have a multitrack/tape input, designed to accept the outputs (returns) from your multitrack recorder. The mic input has a trim pot to control the preamplifier; this increases the very low microphone signal up to line level. The line input has a trim pot as well, but it doesn't have to provide as much amplification. Look at the notations and you'll see zero is in the middle (12:00), with a positive and negative gain on either side. It's really there just to trim incoming levels up or down a bit as needed, but these signals are already at line level. Same goes for the multitrack return, which operates like the line input, but instead feeds the monitor path during tracking.

Phantom, polarity, & high-pass

The microphone preamp section will have buttons for turning phantom power on (+48V), inverting polarity on the incoming signal (ø), and inserting a high-pass filter (⌐) for reducing low frequencies. For example, if you're miking a snare drum on top and bottom, always flip the polarity on the bottom mic to keep the overall snare sound intact.

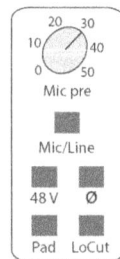

Auxiliary send

Use an auxiliary send, or simply *aux send*, to send a copy of the signal to an external device, such as a signal processor (to add reverb), or to the studio's headphone cue system. Aux sends do not alter the main signal in the channel.

You can always find the aux sends on a console by looking for several pots which are numbered 1-x, depending on how many that board has. It might be as few as two or as many as eight or more. These are configured as either pre- or post-fader, meaning they get their signal source from the audio as it passes before or after the main channel fader. Post-fader aux sends will vary in level as the channel fader is raised or lowered, which works well for reverb during a mix. Pre-fader sends don't change, making them ideal for cue mixes during tracking.

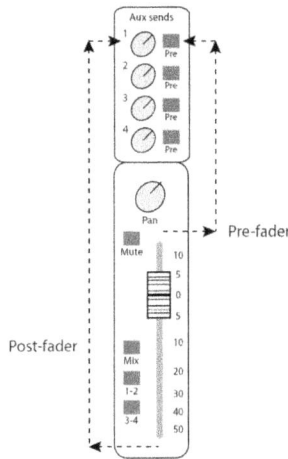

Equalization

Use EQ to adjust the tone of the sound, such as adding treble for more brightness or bass for low end fullness.

You can recognize the EQ section by looking for labels such as high, low, high-mid, and low-mid. If it's a parametric EQ the pots will be grouped, including boost/attenuation, frequency select, and perhaps bandwidth for each band. There's usually also a high-pass filter and possibly a low-pass as well. Some consoles feature the ability to split the EQ between both signal paths, meaning the mids could be used to EQ the mic as it's being recorded while the high and low bands are switched into the monitor path.

Insert point

These are essentially detours on each channel and output bus. You can redirect the signal to an outboard processor, such as a compressor, and then return it back to the same point in the signal flow. The original signal is permanently altered since it's physically rerouted through a processor, which is different from how an aux send works. You find them either on the back of the mixer or on the console's built-in patchbay. Some consoles require you to turn on the insert for each channel, and often you can switch it between channel and monitor path.

Fader

A fader, or *linear potentiometer*, is a vertical slider that controls the overall level of the signal. This affects the strength of the signal being recorded to the multitrack and is also used for balancing the level of your tracks during mixdown. Some consoles have two faders per I/O module, a large and a small one, for the two signal paths (channel and monitor).

Assignment matrix

Use the assignment matrix for routing an input signal to a particular track # on the multitrack recorder. This is a series of buttons labeled 1-8, 1-24, etc. This refers to the number of output buses you have on the console. During mixdown, you would assign each input channel to the main stereo mix bus, which sends it to your 2-track recorder and control room monitors. This might be labeled 2-mix, L-R mix, etc. Some boards automatically route the multitrack returns to the mix bus.

Pan

Use the pan control, or *panoramic potentiometer*, to place that particular track somewhere in the left-right stereo image. This allows you to position a guitar off to the left while moving a backup vocalist over to the right.

Mute

The mute button turns off the channel output. This is useful for muting a vocal track before they begin singing. Some consoles provide a group mute function, which means you can mute several tracks at the same time.

Solo

There are times when it's handy to audition a single track, so instead of muting everything else on the board, just press the solo button. There are two modes: pre-fader and solo-in-place. Pre-fader pulls the signal before the channel fader and pan pot, so the master solo level determines how loud you hear it. Solo-in-place is post-fader, preserving its relative level and stereo placement in the mix.

Output, monitoring, & master functions

Multitrack bus

Multitrack buses connect signals coming through the input channels to selected tracks on the multitrack recorder. Some consoles provide a separate set of faders for these buses, others simply use pots in the master section. During tracking, when you assign a mic signal to a certain numbered output, this is the summing amplifier that receives that signal and sends it on to the recorder. Multiple mic signals can be assigned to the same multitrack bus, such as when recording several vocalists to a single track.

During a mixdown you can use them as sub-groups. For example, you could assign all your drum tracks to one stereo pair of buses. Then, if you need to lower the volume of the drums during the mix, you simply pull these two faders down, rather than grabbing all eight or so drum channel faders. Another option is to set up a stereo drum mix, but keep the original channels assigned to the mix bus. This way you've got the original drums in addition to the stereo drum mix. Put a stereo compressor on this for a parallel compression effect like we described in the mixdown chapter.

Stereo main mix bus

Everything in the channels and aux returns feeds the main mix bus, which is the source for your control room monitors. It's also cabled directly to any 2-track recorders you may be using, which is how the final mix is recorded. Another option for recording your console mix is to connect the mix bus to an audio interface and record into a stereo editor.

Keep this fader at the unity point, which is indicated by either a "U" or "o". Once set, forget about it during the mix and adjust levels with individual channel faders. By the way, when mixing in a DAW you should still set up a stereo mix bus, called a *master fader* in Pro Tools. Everything goes through it, just like on a console, and overall processing such as compression or EQ can be applied. It's also helpful to use its meter to keep an eye on mix levels.

Control room source & monitor level

You can choose what you're listening to in the control room monitors via the monitor source select switches. Options typically include the mix bus, which you use nearly all the time, as well as any external 2-track devices such as a CD player or analog tape recorder. The adjacent volume control only changes how loud the monitor speakers are—it has no affect on any recording levels. So if the pizza guy shows up during a take, you can turn down the monitors while the band plays on.

Studio playback

The console may provide additional outputs to feed a set of speakers in the tracking room. This can be much more convenient for communicating with the musicians as well as playing back a track for them to review. It'll also have source select switches to determine what they hear, such as playing back a control room mix or a demo from your iPhone.

Meters

All consoles provide visual signal strength indicators. There are two main designs: VU and peak. Volume Unit meters will average out signal levels like our ears tend to do. Peak meters show exactly how high a signal transient reached. Don't use the relative volume in the room for setting recording levels—always watch the meters. The idea is to know how high to run signals while avoiding excess noise and distortion.

Oscillator

An oscillator generates test tones or noise that's used for setting levels and checking equipment. With analog tape we'd record sine waves at various frequencies, so when the tape was played back for mixing the engineer could recalibrate their recorder to match how it was first tracked. Running a tone through a device is an easy way to verify operation or calibrate its settings. And we can also add tones creatively to a mix, such as for a hip hop-style kick drum.

Master aux outputs

Auxiliary sends on every channel combine together at the corresponding master aux send, which is a summing amplifier. So, turn up aux three on various channels, then master aux three. Patch this into an effects unit or cue system.

Try to keep auxiliary masters as close to unity as possible, using the individual channel auxes to balance the overall signals. How do you find unity? Sometimes it's straight up at 12:00, other times it might be labelled elsewhere around the dial. And then there are consoles that don't show anything at all unless it's mentioned in the manual.

Aux returns

Auxiliary returns have no connection with aux sends other than the name. They're simply extra inputs to the console that are routed directly to the mix bus. The most common example is adding reverb to a mix. Once you set up your aux sends and route this to the external effects processor, the result has to come back into the console. Connect the output of the effects processor to an aux return, which then sends it along to the mix bus. You don't have to match the aux send # with the aux return #. Some aux returns are mono, others stereo. For returning a stereo effect, such as reverb, two mono returns are required for left and right signals.

Other technical items to know about

Patchbays

Many consoles have integrated patchbays. These are the long rows of connectors that are wired to all the inputs and outputs of your gear in the studio. So instead of crawling behind everything to connect a compressor to channel four, you only have to patch a short cable between the two jacks on the patchbay. If your console doesn't have one, you can easily set one up by purchasing the bays, which can be mounted in a standard equipment rack, and then connecting everything in the back. If you're making your own, there are certain conventions to where things are located. First you need to understand the concept of how jacks relate to each other.

Normalled connections feature two jacks, usually adjacent to each other, which are connected in the back so that the signal coming from one jack will flow into the second without having to patch a cable between them. The reason for this is to facilitate making connections that are used all the time, thereby reducing the amount of patching that must be done. For example, since the multitrack output buses on a console nearly always feed the same numbered tracks on the recorder, studios will normal the bus outputs to their related MT inputs. You still have the two jacks on the patchbay in case you ever need to re-direct somewhere else. Another example is to automatically have certain aux sends feed specific effects devices. If aux four is normalled to your Reverberator XL, then all you have to do is turn up aux four to get signal to the unit without having to patch anything. By patching a cable into either of the two jacks you will break the internal connection, allowing you to redirect the signal.

Half-normalled connections are similar to normalled jacks, except that if you patch into the first of the two jacks the signal will still flow to the other jack. It takes patching cables into both jacks to break the normalled connection. Insert points, used for connecting compressors and noise gates, are always wired half-normalled. Why do this? A couple of reasons, one being that you can use the insert send jack alone to send a copy of that channel's signal somewhere while retaining the original signal flow. The internal connection remains unbroken, and you now have a copy of that signal to do something different with. Another nice use of this system is when you're setting up to record a bass guitar and want to add a compressor. You start by patching the first cable from the insert send to the input of the compressor. During the time it takes to then connect the second cable from the compressor output to the insert return, the original bass signal is still flowing through the channel without interruption. So what? So this means your bass player can keep playing without their cue feed being cut off while you reach for that second cable.

Open patchbay connections feature adjacent jacks that are not internally connected together—they're just individual jacks. An example of this might be the Line In jacks for each channel. They may not be wired to anything, and yet when you want to patch an iPhone or effects output into an input channel, this is where you'd connect it.

Now that you sort of understand this, we can briefly outline typical patchbay layouts. First, you want to line up outputs of gear that normally feed certain inputs all the time. Examples of this include multitrack bus outputs, which should be automatically connected to the inputs of your multitrack recorder. Same goes for the MT outputs, which always return to the same numbered channels on the console. So, MT Bus Out jacks (from the console) will fill one row of your patchbay while the MT Input jacks (to the recorder) will be located on the row directly below. You can do the same with aux sends, where each aux send is automatically wired to a specific effects device input. Channel insert sends and returns must be located on rows immediately above each other since they're half-normalled. That's enough for now—there are several available books on wiring studios that can help show the details on patchbay design.

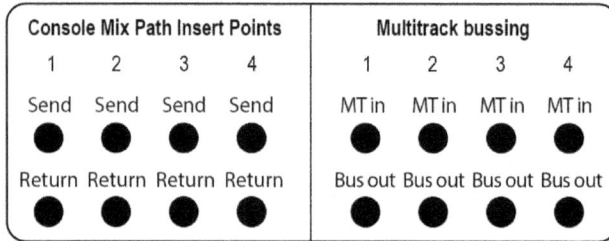

Console Mix Path Insert Points				Multitrack bussing			
1	2	3	4	1	2	3	4
Send	Send	Send	Send	MT in	MT in	MT in	MT in
●	●	●	●	●	●	●	●
Return	Return	Return	Return	Bus out	Bus out	Bus out	Bus out
●	●	●	●	●	●	●	●

Automation

Mixing in the old days was always an experience. Every little change during the song, such as fader movements, EQ adjustments, or turning an effect on and off at a specific time had to be done by hand. We'd put pieces of tape all over the board showing exactly where the faders and pots should be set, and then we'd assign specific moves for each of us. And then someone figured out how to get the faders to move on their own, following the programmed level changes we'd set up. With analog consoles this is called VCA automation (voltage controlled amplifier); the computer sends data to the console that is then translated into low-voltage signals. This in turn controls the fader amplifiers. Solid State Logic (SSL) took this to the extreme when they released consoles featuring Total Recall, where nearly everything on the board could be saved and recalled in an instant.

Digital consoles have all this built-in, of course, including motorized faders. The ability to recall complete mix settings, including what you get in a DAW, makes it so much easier to revisit a mix later when the client points out the chorus where the lead vocal somehow disappeared...

Digital vs analog

Yes, companies still build analog consoles, big and small. Part of the reason is for the sound you get from the circuitry in an analog console, but as mentioned earlier some people use other gear to get "that analog sound" and do everything else in the computer or on a digital board. Analog consoles are somewhat easier to learn because there are dedicated controls for every function and for every channel. Digital consoles, however, will condense these physical controls so that fewer switches and faders control everything. So, you might have a central section that controls EQ, compression, and effects, but you need to select a specific channel to work on. One of the great benefits of digital is the built-in signal processing (EQ, compression, gating, and effects). You can also save all your settings with numerous scenes for easy recall later. These consoles usually integrate well with computer-based recording systems.

Here is a comparison between an analog console and what it might look like as a digital design. You'll notice fewer controls on the digital model, relying instead on a pod of physical pots that control whatever you've selected in the menu. Want to adjust the EQ on channel five? Select channel five, then twirl the EQ dials.

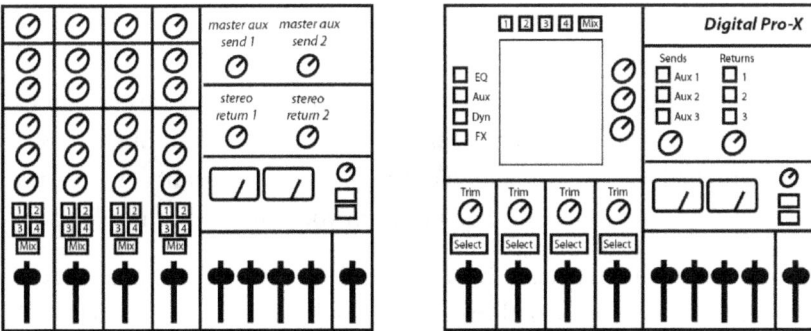

Which is better? It depends on the quality of the unit, how it's designed, the user interface, and the features that are important to you. Digital boards have largely taken over for live sound because they sound great and provide much more functionality and convenience than running analog boards with tons of outboard gear. In the studio, though, it really depends on your situation, budget, and priorities.

I'd like to buy a console—where do I start?

If you're looking to buy an analog console, there are a few issues to consider and understand. The less expensive recording consoles are pretty good these days. They usually have eight multitrack buses, and they're physically small so they don't take up much room. They have most of the normal features you need such as parametric EQ, monitor path or monitor capability for tracking, plenty of aux sends, etc. Aside from the limited number of buses, the biggest drawback is in the quality of the physical controls and electronic components as well as the fact that multiple channels are combined into a single circuit board design. This saves lots of money in the price of the board, but means that if one channel breaks you have to pull several channels (or the entire board) out for repair. These consoles cost less than $10k and are generally great for the money.

On the other end of the spectrum are the really large consoles that may cost $100k or more. They feature well-designed circuitry and controls, extra features, and individual module design so you can replace one channel at a time. Very nice, but very expensive. These have trended down in recent years, although you will always find a use for them in some high-end facilities. There are also some analog consoles priced in between the low and high end, featuring a hybrid of quality and features that are worth investigating depending on your budget and needs.

Some manufacturers make hybrid consoles that combine analog audio circuitry with digital signal processing, including integration with software systems. This provides the best of two worlds—hardware control over your DAW with the sonic quality of analog circuits.

Digital recording consoles are also available in a wide range of price points; they all feature the general digital capabilities described earlier. They take up much less space in the room and you don't have to buy stacks of outboard equipment such as compressors and effects. Much of the difference lies in the quality of the converters. These components convert analog audio to digital and vice versa and are critical for a quality sound. Cheap devices have cheap converters, so buy as much quality as you can afford.

The final option is to forget the console entirely. Run a DAW on your computer, plug the mics into an audio interface (which has the preamps), and use a hardware controller that uses traditional faders and pots to control your DAW. Keep in mind that the preamps are the most crucial component for audio quality and flavor, so get really good ones.

Lastly, keep in mind that recording consoles are different from live reinforcement mixers. Recording requires the second monitor path capability we discussed earlier, and live boards do not have this. You can set up a simple live mixer to do some recording, but it's much easier and more powerful with a recording console.

ELEVEN

MICROPHONE DESIGN

All microphones are *transducers*, which simply means they are devices that transfer energy from one system to another. Microphones take acoustic sound waves and generate electrical signals. Other transducers in audio include tape recorders, speakers, guitar pickups, and phono cartridges for record players.

Although the sole purpose of a microphone is to capture an acoustic event and convert it to an electrical signal, there are multiple ways to achieve this. None is particularly any better than another, and variation in microphone design is deliberate so as to provide each model with a distinct sound. The result is an artist's palette of mics that provide different nuances to your recordings, so you can match just the right mic to that special soprano in your life.

Audio example 70: Comparing mics using a pair of Shure KSM32 vs AKG 414

Audio example 71: Comparing Shure SM57, Sennheiser e906, & Rode NT1000

The two primary design types are *dynamic* and *condenser*. Other variables such as component materials and circuit design will also contribute to each mic's distinctive sound.

Dynamic microphones

This design employs the physics principle of electromagnetic induction. Inside the capsule of the mic a coil of wire is suspended within a magnetic field. The microphone diaphragm is attached to the coil, and when sound waves hit and move the diaphragm, the coil of wire also moves back and forth within the magnetic field. This generates an output voltage that varies according to the diaphragm movement. So, the output signal varies analogous to the changes in the original acoustic sound wave...yes, that's where we get the term *analog*.

Examples of dynamic microphones include the Shure SM57, Shure SM58, Sennheiser MD421, and Audix D4.

Two variations of dynamic microphones

Moving Coil: This is the most common type of dynamic microphone; probably most of the dynamics in your collection are moving coil.

Ribbon: Instead of a coil of wire, a thin ribbon of electrically conductive material is suspended within a magnetic field. The principle is the same as for moving coil, although since sound can strike the ribbon on both sides, these mics are inherently bi-directional. These are very popular for studio recording due to their warm, smooth response.

Condenser microphones

The principle of electrostatics is the basis for condenser mics, which are also known as capacitor microphones. These have two electrically charged parallel plates to transduce acoustic waves; one is movable (the diaphragm) and the other is fixed, effectively forming a capacitor. When sound waves hit the movable plate, the distance between the two plates varies. This changes the capacitance, generating a corresponding output current.

Condenser mics require a DC power source to charge the plates and power an internal preamp; this low-level amplifier reduces high internal impedance and increases the output level a bit. This is not the same as the mic preamp on the console that boosts the mic signal up to standard line level. The power source is called phantom power (+48V) and is usually supplied from the console through the mic cable. Look for a switch on each channel, or perhaps for a bank of channels on the back of the console. If you're using an audio interface, it'll provide a phantom power switch. If you plug in a condenser mic and nothing happens, verify that phantom is on for that channel (mute the channel before turning phantom on/off).

Some condenser mics you might run into include the Shure SM-81, Audio Technica AT-4040, and the Neumann TLM series.

Signal levels in a studio

There are different strata of signal levels in a recording studio, from AC power (highest) down to microphones (lowest). Most of the equipment, including the console, operates somewhere in the middle, known as *line level*. Since the output signal of a microphone is significantly lower than this, we have to run it through a special amplifier that increases it to line level. These are called *microphone preamplifiers*; every console channel has a preamp, your audio interface will have at least one or more, and you can buy external preamps that are then connected directly to your recorder. Due to the large amount of signal gain required, the amp introduces a bit of coloration to the sound. We can use that to great advantage, selecting different preamps for each miking situation. Just like the fact that microphones are intentionally designed with their own inherent sound, mic preamps come in a vast array of flavors. Quality does count, however, and cheap preamps will sound, well, cheap. They don't provide enough gain to power mics adequately, they sound brittle, thin, and, well, cheap. If you're beginning to build your equipment collection, splurge on the mics and preamps and it'll pay off.

Directionality

There are times when sound needs to be picked up from only one direction, other times from all around the microphone. Mics are designed to be selective from where they collect sound waves. We call these directional characteristics *polar patterns*, and there are three primary types:

- Uni-directional (cardioid)
- Bi-directional
- Omni-directional

A cardioid microphone will collect sound primarily from on-axis, meaning the front of the microphone. This is useful when you don't want a particular mic to pick up other sounds that are intended for other mics. However, these do not completely disregard sounds from off-axis. As you get farther away from on-axis (moving toward the side and back of the mic), the microphone gradually attenuates the sound it picks up. This results not only in a signal level reduction, but also causes changes in the frequency response of the sound, altering its tone. This means that the sound will be unnatural as you get more off-axis; the term is *off-axis coloration*, and it's not something normally desired in your recording.

Bi-directional microphones pick up sound from in front and in back of the mic, but reject sounds from each side. Omnidirectional mics accept sound waves from all around, though there is a slight directional characteristic for high frequencies, resulting in a slight roll-off from off-axis of the microphone.

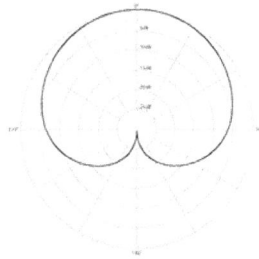

Cardioid pickup

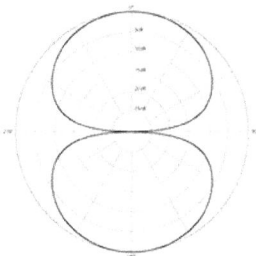

Bi-directional pickup   Omni-directional pickup

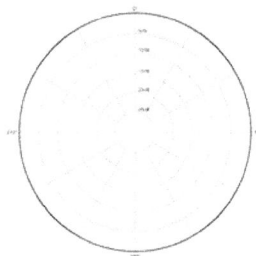

To achieve directionality, microphones have ports (openings or slots) along the outside of the casing. These ports lead to internal chambers that are designed to delay sound waves before striking the diaphragm from the inside. Cancellation is achieved when a soundwave coming from behind the microphone enters these ports while also diffracting (bending) around to the front of the mic to strike the diaphragm from the outside. When these two identical sound waves hit the diaphragm at the same time from opposite directions, the result is cancellation. Back to physics class, if two people push against each other with the same force, the result will be...zero. Nobody moves. Now, with audio signals it won't be total cancellation, depending on relative position to the mic and the fact that all frequencies have different wavelengths. This will be discussed further in the acoustics chapter, but the idea is that different frequencies will be attenuated/cancelled as the source moves around the microphone, resulting in a colored, altered sound. The lesson here? Listen carefully to each of your mics to hear what they're picking up; you definitely don't want to discover this during mixdown when it's too late to fix it. You might need to space your musicians away from each other a bit more, put up a gobo in between, or even decide to overdub a part later.

Shotgun microphones take the off-axis cancellation approach to the extreme. On-axis sounds enter with little attenuation, but sounds entering progressively off-axis are severely attenuated, particularly higher frequencies. These mics are intended for very tightly focused aiming, such as actors filming on set (or eavesdropping across the football field).

A common alternative design for implementing polar patterns involves dual capsules (diaphragms), one being omni-directional and the other bi-directional. These two elements are mathematically summed together as needed to achieve any of the three primary patterns as well as variations on these (hyper-cardioid, super-cardioid, etc). These microphones will have a switch for selecting the desired pattern. Many large, dual-capsule condenser microphones are on the market today.

Frequency Response

As audio passes through any device or system, the relative balance of low, mid, and high frequencies is affected. Ideally we usually want a device to keep the original signal intact with no change. Microphones are a different story, however. I mentioned a few moments ago that microphones are intentionally designed to produce unique sound nuances. Microphones with large diaphragms tend to reproduce low frequencies better, making them good candidates for kick drums, floor toms, and upright bass. Small diaphragm mics are good at high frequencies, so they can be ideal for cymbals. Some microphones feature very flat response and are intended to remain neutral to the sound. Others, like the SM57 and 58, have a frequency response bump that adds energy in certain regions. There are no rules, but these characteristics can provide some guidance when selecting a mic. When you buy a microphone, it comes with a frequency response chart to show how it passes sound, so check it out and compare with other mics to see how they respond differently.

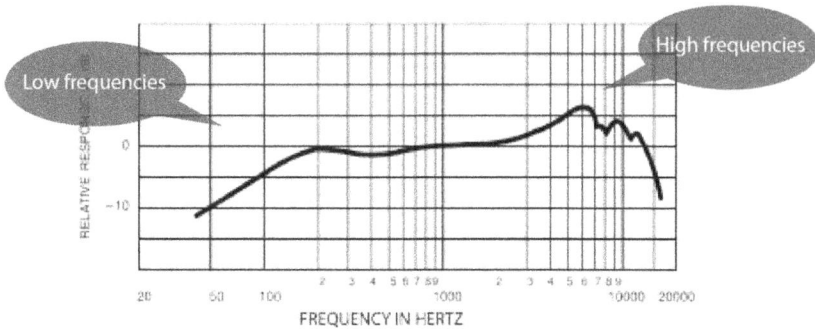

Something to be aware of is that when using cardioid or bi-directional microphones very close to a sound source, low frequencies are over-emphasized. This low-end boost is known as *proximity effect*. Sometimes this is desirable, such as for giving radio DJs that characteristic deep, booming voice. Mostly it just muddies your sound and should be reduced as much as possible. There are a few ways to accomplish this, such as moving the mic away a bit, using an omnidirectional mic instead, or by switching in a low-cut filter. Many microphones have a switch that attenuates low frequencies, usually around 75Hz or 80Hz, and your console will also have these on each channel. Digital consoles and plug-in EQs allow you to dial in the frequency you want to start attenuating, so it could be set higher for a female vocal and lower for an acoustic guitar or piano.

Transient Response

How fast and accurately does the diaphragm respond when an acoustic wave strikes it? It takes time for the diaphragm to move and for the electrical signal to be generated. Audio transients are very brief bursts of signal found in all sounds, especially percussive events such as drums, guitar strumming, even piano chords. They are often so short many microphones simply ignore them. High frequencies have very short wavelengths and fluctuate very quickly. The quicker the response of the microphone, the more accurate the reproduction. Condenser microphones exhibit a faster, more accurate response that reproduces clear high frequencies, whereas dynamic mics, particularly ribbons, are more sluggish and tend to round off the waveform, resulting in a smoother, more mellow sound. Neither of these is better than the other—they're just different. So, if you place a condenser mic on a trumpet and it peels your ears back, replace it with a dynamic and enjoy the mellower sound. If you're looking for a crisp, bright sound from your cymbals, put a condenser on them.

Audio example 72: Comparing dynamic vs condenser mics

Sensitivity

This specification indicates a particular microphone's capability for output voltage level. A higher rating provides more output signal, which means you don't have to turn up the mic preamplifier on the console as much. This is beneficial because less gain means less noise and a cleaner signal. You'll notice this particularly when comparing condensers, dynamics and especially ribbons, all of which exhibit very different output levels (condensers are hotter primarily due to their internal preamp, but the idea remains the same).

Overload

Just like with any audio device, there is a limit of how much signal level a microphone can handle. A mic is overloaded when the SPL level (sound pressure level) is so high it distorts (clips) the diaphragm and/or electronics. High level sources such as drums or brass can potentially cause problems, though most microphones are pretty robust. Dynamic mics feature a higher overload tolerance than do condensers. If you're distorting the mic itself (as opposed to overloading the mic input on the console), either move the mic farther away or use an attenuation pad if the mic has one. I once recorded a vocalist who had such a powerful voice she was overloading the mic—and it was a very nice, very expensive condenser. The mic pad did the trick and we got a great sound.

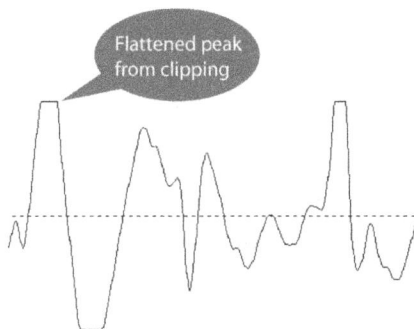

Equivalent noise

This is a rather obscure specification that keeps very few engineers up at night. All electronics make a small amount of noise as electrons move around doing their thing. This self-noise is quite minor compared to other noise sources in the recording chain, but becomes a bit more important with digital recording equipment as the noise floor is much lower than with analog recording. Any professional-quality microphone is sufficient, so don't spend your time trying to measure it.

Impedance

This little-understood term is actually quite important in the design and connections for audio equipment. For now we'll just leave it as the mic's ability to provide a certain signal "strength" as compared to what the console or audio interface is asking for in order to obtain optimum transfer of signal. Remember when you go to a concert and they have restricted gates and entrances to control the crowd going in? Think about if they either closed these nearly shut or opened everything completely. People would pile up trying to get in or you'd get a stampede. You get the idea. Maybe.

In more practical terms, all professional microphones (and professional audio equipment in general) are low impedance (low-Z), so don't use high-Z mics, which usually have 1/4" connectors and act as radio antennas for any available broadcast that happens to be floating through the room. Low-Z mics are much better at preventing outside interference and extraneous noise (motors, fluorescent lights, radios). High-Z mics also suffer from high-frequency loss over distance.

Some microphone preamplifiers let you select different impedance values within the low-Z range. So what? This slightly changes the tone of the sound you're getting from the preamp, which provides a really nice, subtle nuance for your recording.

Balanced microphone cables

Professional low-impedance (low-Z) microphones use cables employing two signal-carrying wires in addition to a ground wire (shield). These signal wires are twisted around each other throughout the cable, and the shield is most often braided around the two wires. This provides maximum protection from outside noise interference, or RF (radio frequency).

How does it do this? Audio signals are AC current, meaning they alternate positive/negative between the two signal wires. Any outside interference leaks into the cable as a common polarity DC signal. When it arrives at the end it's cancelled out because all balanced audio gear (professional equipment) is designed to accept AC signals only. The shield in the cable drains extraneous noise by shunting it to ground.

Balanced audio requires a connector with three points, so we use either XLR connectors or TRS 1/4" (tip, ring, sleeve), and the three pins are numbered so as to match on each end of the cable. Pin 1 is always the shield (ground); pins 2 and 3 carry the alternating polarity audio signal.

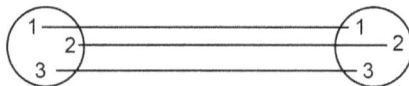

Phantom power

As we described a bit earlier, condenser microphones require an external power source, usually 48 volts DC. This is sent from the console through the mic cable to the microphone, but does not damage the mic (or anything else it's plugged into). Audio equipment works with alternating current and looks for the difference between the two wires; phantom power is DC. Since there is no difference at the input it is ignored by the device, meaning it won't damage anything plugged into that channel. So it's not a huge deal to leave phantom power turned on, but if the console has individual channel phantom switches I leave them off unless a condenser is plugged in. Make sure you don't turn phantom on or off while the channel is on and routed through the console. It'll send a spike through the system that you'll hear in the monitors and potentially in the cue headphones.

PZM (Pressure Zone Microphone)

The generic term for *PZM* (a product trademark of Crown, Inc.) is *boundary microphone*, and refers to a microphone where the mic element (diaphragm assembly) is mounted on a flat plate. The concept is to reduce phase cancellations that occur when using a traditional microphone stand. A mic on a stand is elevated above floor level by several feet. The sound reaches the mic, but also bounces off the floor and into the mic somewhat later than the original wave. Back to acoustics, when two identical waveforms arrive at different times, phase cancellation occurs, which means the tone of the sound is altered in a negative way. With a boundary mic, there is no phase-altering reflection from the surface since the mic is directly mounted on that surface; all it gets are direct soundwaves from the source.

PZMs are popular for miking underneath a grand piano lid, on the floor, or on a wall. Try placing a close mic on a guitar cabinet, then adding a PZM on the floor several feet away to capture the overall room sound.

Don't mount these on a traditional mic stand as they're designed to be placed on a large surface.

Tube microphones

I mentioned earlier that condenser microphones require power to amplify the internal low-level signal. This amplifier can be either a FET amp (solid state) or a small vacuum tube. As in all things audio, tubes are desirable as they introduce color to a sound, specifically adding 3rd harmonic distortion into the signal path. This translates into a warmth and smoothness that engineers and musicians really like, so there are many tube microphones on the market these days.

Microphone switches

Pay attention to the various switches on your mics and make sure they're set to what you want. Polar patterns are typically set to cardioid (front pickup only), and filters and attenuation switches are off until needed. Not all microphones have these switches, so if you don't see a polar pattern switch on your SM57, don't tear it apart looking for one.

- Polar pattern select (condensers)
- Attenuation pad to reduce incoming signal level (condensers)
- Low-cut filter to attenuate low frequency sounds (condensers & dynamics)

Direct box

These devices connect electronic instruments such as keyboards or guitars to the console or audio interface. They're available in either passive or active models. An active DI requires phantom or battery power and uses active electronics to provide a "hotter", perhaps more aggressive output. You can find very nice passive models as well; the point is to buy good ones—the cheap models sound terrible. A decent DI can be had for $100 or so, such as from Radial or Nady, while the really nice ones go for several hundred dollars.

Direct boxes have a 1/4" input jack, where you plug in the guitar, and an XLR output jack where you connect a standard mic cable to the console or interface mic input. Once you plug everything in and turn on the channel, listen for a hum (ground loop). If you hear it, flip the ground lift switch on the box. It doesn't matter which setting you use, just pick the position that reduces the noise.

Some DIs also have an instrument/amp switch, which sets the expected incoming signal level. If you're plugging in a guitar or other instrument, set this to "instrument", which is lower. You can, however, take the output of a guitar or keyboard amplifier, which has a much higher signal level, and run that through the box. In this situation set the switch to "amp".

And in case you're wondering, "DI" stands for "direct injection", meaning the box takes your signal directly into the system rather than through a microphone. Nobody says "direct injection", though, so I wouldn't try it out on your friends. They'll just stare at you.

Direct boxes perform the following three functions:

- Reduces output level of the instrument (roughly line-level) to mic level so it can feed a mic preamplifier.
- Changes the instrument's high-Z output (unbalanced line) to a low-Z source (balanced) as needed for the mic input.
- Isolates audio signal, eliminating ground loop hum.

Microphone accessories

Windscreen

Also known as a pop filter. Hold your hand in front of your mouth and say something with lots of "p" and "b" sounds. Feel the puff of air? This can overload the mic diaphragm, causing those tremendous booms you heard in the school auditorium when your principal was lecturing to the student body. Another option is to place the mic slightly above the mouth, angled downward a bit. The air stream goes down, not up, so this can also take care of it. Don't use the thick foam covers; they're meant for outdoor wind and attenuate too much high frequency information.

Shock mount

Stand in a room with a drummer and feel the floor vibrations, or listen to the thud when a lecturer accidentally kicks the podium. Shock mounts are designed to isolate the microphone from vibrations that are transmitted from the stand into the mic. They have an elastic strap which functions like a shock absorber, holding the mic so it doesn't physically touch the plastic or metal of the mount.

Stand and boom

The boom is the horizontal extender attached to the vertical stand. And you were worried about that one.

Stereo stand adapter

This gadget mounts two microphones on a single stand, most often in a specific stereo miking configuration. They come in all sorts of designs and configurations, but the application is the same.

TWELVE

MICROPHONE TECHNIQUE

Understanding the design and characteristics of microphones gives us a foundation for making better choices during a session. But the question remains: How do I use these things and get a great sound? Microphone technique is very subjective; there are a few guidelines, but much is up to the individual, current trends, and the particular situation. The rule is not to worry about rules too much and just experiment. Sure, I'll give you a couple of things to keep in mind that might cause problems, but generally you should just try lots of things to see what you like.

The first thing I'll tell you is to go listen to the instrument in the room before you pick up a microphone. What does it sound like? Walk around, move your head, ask the musician where the sweet spots are as all instruments have different radiating characteristics (where the sound goes). Experiment with moving the musician to different locations in the room. Once you find a spot you like, go pick a mic that best matches that sound. Not sure? No problem, just try a few. Too many young engineers throw up a mic and start turning EQ dials. Listen first, find the right mic, then find the right place for that mic. Always.

Placement considerations

It helps to understand some basic acoustics as to what happens when a sound radiates from a source into the room. With this in mind you can consider placement options that give you a certain desired sound, such as close and tight or farther back in a room with more reverberation.

When a sound is generated in a space, it goes through three stages: direct, early reflections, and reverberation. The direct sound comes straight from the source into the microphone. At the same time it begins radiating outward into the room, reflecting back from nearby boundaries (walls, ceiling, windows). These are called early reflections and give us a sense of a room's size. As the sound continues bouncing around the room it gradually turns into a dense field of reflections we call reverberation. Though all spaces follow this principle, the nature of the room dictates how it translates into what you actually hear. A large room with hard surfaces will generate a fair amount of reverb, whereas in a small room with furniture or other soft materials it will be negligible.

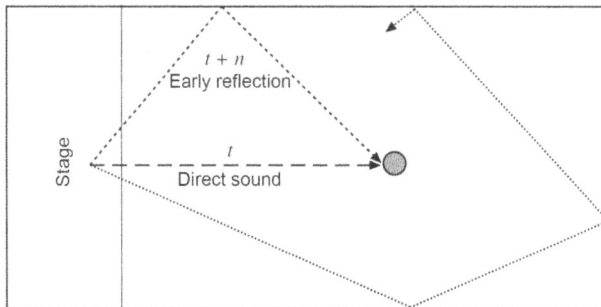

Placing a microphone close to an instrument, say less than a few feet, will pick up direct sound with little reflections and certainly no reverb. As you move the mic farther away, it begins detecting reflections from the walls or ceiling—you start to hear the "room" as part of the sound. Place it far away in the room and it'll pick up mostly, if not exclusively, reverberation. With this in mind, let's break these down into categories so we can explain what's involved with each.

Close miking

Close miking is intended to capture the immediate sound of the source with little or no room reflections. The idea is that if we get a clean sound of the instrument, we can add artificial reverb later in the mix to provide a sense of space and place. The distances involved depend on the sound source and acoustics of the room, but generally we're talking anywhere from an inch to a few feet or more.

Experiment with slightly different distances from the source. For example, a kick drum sounds very different when mic'd near the beater head vs farther back in the shell. Place a mic on a mounted tom about an inch away, move it back to six inches, and take a listen. A couple of inches either way on a guitar cabinet can make a significant difference. Take some time to play with it; it's far better to get the sound you want from the mic rather than trying to EQ it to death.

Placing a mic too close, though, results in a colored tonal quality—it doesn't sound natural. A good example is the piano, which is large, complex, and needs space and time for everything to radiate and merge. So, while we could place a mic 6" directly over one section of the strings, you're not going to get the entire frequency range of the instrument; the result is an uneven piano sound. Miking very close with directional microphones also results in a low-frequency boost called proximity effect, which makes it sound boomy and muddy. In this case simply move the mic a bit farther away or use an omnidirectional model.

Close miking issues

- Doesn't sound full or natural? Try backing the mic away and experiment with placement. Small changes can make a big difference.
- If it sounds too "in your face", back the mic off a tad.
- Boomy sound? Pull the mic away a bit, turn on the low-cut filter, or use an omni.

- Muddy? Move the mic around a bit, attenuate the low-mids (see the section on EQ).
- Other instruments coming into the mic? Move the musicians farther away from each other in the room.
- Still doesn't sound like the guitar in the room? Try a different mic.

Distant miking

For distant miking, the idea is to get a more rounded sound from your source blended with some room ambiance (reverb). This could be a balance between direct sound and reverberation, or it could be reverb only. The microphone might be located anywhere from a few feet or all the way across the room. It's also common to combine a distant microphone with a close mic, capturing both the direct and room sounds.

So we might put a microphone several feet away from an ensemble, adjusting the distance to achieve a desired balance between direct and room. Or we can move that mic farther back in the reverberant field strictly to capture the acoustics of the space. Similarly, we set up a second mic for a guitar cabinet, but place it far back in the room strictly for the reverb. Although we generally think of close-miking drums in the studio, lots of great records have been made with drum mics placed several feet away from the kit. This takes advantage of having a nice, large room with high ceilings and results in a really big sound (Phil Collins' *In the Air Tonight* is a great example).

How much of the room sound do you want? This depends on the track itself and how it fits in the song, but also on the acoustical characteristics of the room. A drywall room sounds quite different from plaster or wood. Wood floors generally are more desirable than carpet. If the walls and ceiling are flat with nothing to break up the contour, hard reflections are returned to the mic and cause problems. If your room doesn't sound so great, you can move your mics closer or treat the walls with acoustic materials. If you're blessed with a wonderful sounding, large, lively space, take advantage of it and add more distant miking to your recordings.

Audio example 73: Close vs distant miking

Miking issues to watch for

Here are a few things you should be aware of when setting up your mics. Some of these you'll easily hear, but others will silently cause trouble unless you're aware they exist.

Leakage

Leakage is when a microphone picks up sound from something other than its intended source. If you have individual vocalists in the room, each with their own mic and standing fairly close together, each mic will pick up the other singers off-axis, which sounds very different. Leakage colors both the leaked-into mic as well as the mic focused on the other source. The snare might be picked up off-axis by the kick mic; when you're mixing the song and work for that perfect snare sound, you then turn on the kick track and wonder what happened to the snare. If the leakage is bad enough it becomes difficult to work on individual track sounds.

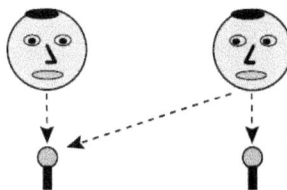

Listen carefully to each of your tracks while you are setting up for recording so you don't run into huge problems later. Your goal is to have control over the level and tone of each track, so reducing leakage is a major objective. Distance can be your friend here. The farther away a sound is from a mic, the less it will be picked up. Every time you double the distance between the mic and a sound source, the signal level drops around 4–6dB. So if you place a mic about 6 inches from an instrument, and other instruments are several yards away, you can do the math and see how effective this can be. It's not perfect, though there are additional tricks that can help:

- Position mics closer to their sources.
- Use cardioid mics and make sure other sources are located off-axis.
- Position gobos (movable partitions) between sources.
- Use different rooms, such as an isolation booth.
- Overdub later.

Having said all of this, there are many situations where leakage is actually desirable. A big band in the studio has a rich, big sound due to everything playing together in the same space. A jazz combo tends to lean toward a raw, open sound between members of the group. Phil Spector's famous "wall of sound" in the studio came from having as many musicians as he could fit in the room; since sound travels relatively slowly, everything gradually bounced around and blended together into a huge wash of sound. The early decades of recording featured musicians in a large room playing together, creating a common vibe and musical connection that's generally missing with over-dubbed records. So again, it all goes back to whatever sounds good for the situation.

Boundary interference & phasing

Sounds radiating from a source not only travel directly to the mic, but also to surrounding surfaces such as walls, floor, and ceiling. These reflected waves can cause phase cancellation and attenuation at certain frequencies. Say you're miking a guitar and it's located near a wall. The guitar sound goes directly into the mic, but it also strikes the nearby wall surface and then reflects back into

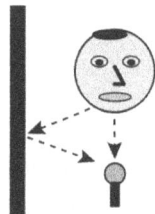

the mic. Since it takes longer for the reflection to arrive, you've got two copies of the guitar signal that are out of time with each other. They're combined at the microphone, resulting in an altered timbre.

The same goes for a mic on a stand several feet above the floor. Usually the mic is placed fairly close to the source, but at increasing distances you'll run into floor reflections. Since frequencies have different wavelengths the phase shift varies throughout the audible spectrum. The result is a succession of dips and peaks in the frequency response (constructive and destructive interference). We call this a *comb-filter* effect. It sounds very colored in tonality, but it usually takes practice to learn to detect it. If you remember sine waves from school (sorry about that), think of two identical sine waves. Now slide one of them over slightly—the peaks and troughs don't quite line up anymore. When you add them algebraically (another apology here), you don't have a sine wave anymore. This will sound different from the original, usually in a way that you won't like.

Placing two microphones on a source can cause the same problem. If they're not exactly the same distance from the source, you've got two identical sounds traveling different path lengths and being summed at the mic. Either ensure both capsules are at the same distance or follow the 3:1 rule. Religiously. Engineers tend to ignore all kinds of recommendations for audio, but they nearly always follow this one. For every one unit of distance between each mic and its source, the distance between microphones needs to be at least three times that mic-to-source distance. As a simple example, if the first mic is six inches from the guitar, place the second microphone at least eighteen inches away from the first mic.

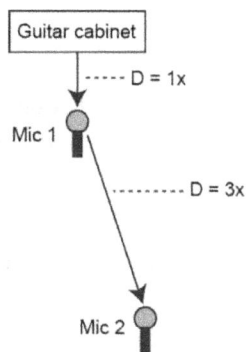

Audio example 74: Guitar mics phasing

Drums are the biggest challenge when it comes to phasing since you typically have so many in a small area. Turn one mic on, then another, and then another until you hear a weird tonal change. This is where you probably have some phasing issues. For example, listen to the snare mic alone, then add others into the mix. At some point your snare sound will change, meaning either you've got leakage (you now also hear the snare from the overhead or tom mic) or phasing (from multiple mics picking up the snare at different distances). Try flipping the polarity switch on the last mic you turned on and see if it helps. Sometimes you'll need to adjust mic positions a bit around the kit.

Audio example 75: Adding individual drum mics on the set to hear how each part changes. Listen how the snare sounds different when toms and OH mics are on.

So what exactly does it sound like? *Ethereal* and *hollow* are the best I can come up with to describe phasing. The most obvious phase issue is when a speaker is wired incorrectly. Route something with lots of bass into two channels of your console, flipping the polarity switch on one channel only. Now pan them center and you'll hear the bass pretty much disappear. Another way is to take a track in your DAW, duplicate it, then nudge one of them over very slightly. You'll hear it immediately. If you instead put a delay on one of these tracks, you can adjust delay time (between 0 and 16ms) and get a sense of how changing distance alters the result.

Audio example 76: Voice phasing

Audio example 77: Drums phasing

How to reduce phasing issues

- Keep mics close to the source.
- Stay away from walls, windows, and other reflective surfaces.
- Use only one mic.
- Follow the 3:1 rule for multiple mics.
- Acoustically treat walls to reduce direct reflections.

Sibilance & pops

Sibilance is excessive energy around 6kHz. We most often think of this when miking vocals, as the "s" consonant is particularly troublesome in this range. But we can have the same issue with squeaking guitar strings. A pop occurs when too much air strikes the mic diaphragm, overloading it. The consonants "p" and "b" are the usual culprits; hold your hand in front of your mouth and say something with these sounds and you'll feel it. The best defense against sibilance and pops is mic placement. Position it just above the mouth, perhaps pointing down a bit. Since the airflow is generally downward, this usually avoids the problem. You can also put up a pop filter in front of the mic; these are thin nylon or metal mesh screens that help diffuse sound pressure without affecting the sound.

A de-esser plugin can be effective during mixdown to reduce sibilance. Adjust the frequency range for the situation and set a threshold so it doesn't kick in too much. The old way to de-ess a vocal track was to run a copy into an external EQ, crank up the 6kHz range considerably in a big frown-face curve, then run this as a side-chain into a compressor that's inserted into the main vocal track. The compressor will only attenuate when this extra "s" energy comes in the sidechain. (We'll explain compressors and sidechains later in the book.)

If you've still got a low-frequency pop on a track, try a low-cut filter; the danger is that it'll probably negatively affect your sound, so adjust it the best you can for a compromise.

Audio example 78: Sibilance

Audio example 79: Removing a vocal pop with a low-cut filter

Starting points for miking

Here are a few ideas to get you started. Remember the issues we've been discussing, experiment with different mics and placements, and enjoy the journey.

Drum set

There are different schools of thought regarding drum miking. Some engineers go the minimalist route, with only two or three microphones on the entire kit. At the other end you'll see individual mics on everything. Each approach gives you a different sound, so it depends on the project. Generally dynamic mics are most common for drums, with condensers used for cymbals and hi-hat because they are brighter and clearer in the high frequency range. However, many engineers prefer large diaphragm condensers for drums, especially toms and kick. Make sure your mics are out of the way of flying drum sticks; when working with non-professionals it's usually a compromise between finding a good spot for sound, yet staying out of the drummer's way.

Overheads/cymbals: A pair of condensers or ribbons over a drum set is ideal for capturing not only the cymbals, but most of the entire kit. In the early days this was pretty much it before engineers began adding separate mics on the kick, then snare, and so on. Start a few feet above the cymbals in either a spaced pair or XY configuration; if they sound too harsh move them

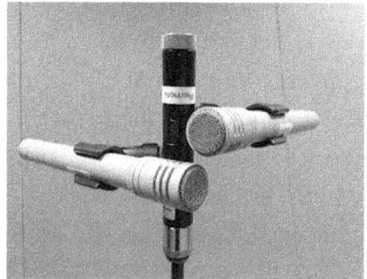

higher. Try to keep both mics equidistant from the snare to preserve a solid stereo image.

The other trick for overheads is a bit counter-intuitive than what you see all the time. Most engineers place the mics on the left and right sides of the kit directly over the ride and crash cymbals. The issue is that snare and kick are usually panned center in a mix, so it's important to capture these exactly in the middle from the overheads. The intersection of these two instruments is off-angle to the cymbal lineup, so we need to swing the overheads clockwise a bit so they're perpendicular to the kick and snare, maintaining the same distance to the snare.

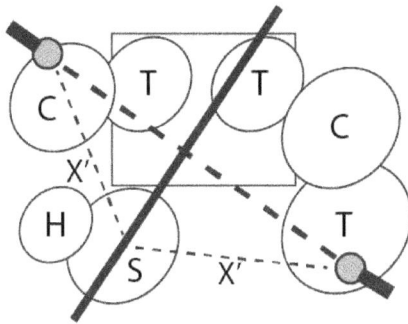

High hat: For the high-hat, point a small-diaphragm condenser several inches above the top hat's outer rim facing away from the snare (to decrease snare leakage). Don't face it directly into the side where the two cymbals clash together; if you don't believe me, go stick your face in that spot and wait for it.

Kick drum: Use a large diaphragm mic like the Shure Beta 52 or AKG D112. These are good for rich, punchy low frequencies. Place the mic inside the shell facing toward the drummer. For a sharper attack, put it deeper inside and aim it where the beater hits. To get a deeper, punchier sound pull the mic back toward the front of the shell. Most engineers like a front head with a hole; it's not preferable to remove the head completely as you lose low end, high end, and tightness. Often you'll see a mic placed just at the edge of the hole; experiment with different placements and you'll get very different sounds. Stuffing blankets, sweatshirts, or jelly doughnuts up against the head inside the drum dampens the resonance, giving a tighter, more controlled sound.

Audio example 80: Kick drum mic close to the beater, then farther back

---

Snare drum: The Shure SM57 is a long-standing favorite. You want a mic that can capture that crisp attack of a snare, and the 57 has a bump in the high-mid frequency range that's perfect for this. Place it over the edge of the snare, maybe two inches inside the rim, and a couple inches high.

Mid-toms: Similar to the snare, you want to emphasize the attack sound of the tom, where the drumstick whacks it. The 57 works fine, but it begins losing lower frequencies that larger toms have. For a fuller tom sound try the Sennheiser 421 or a large diaphragm condenser. Place them like we did for the snare, but a little higher.

Floor tom: Put a large diaphragm dynamic or condenser mic a few inches inside the rim and around 4–6 inches high; too close and it'll sound boomy and unnatural. Giving it some space results in a bigger sound.

Check out the wear patterns on the drum heads. If they're tightly contained in a small spot, you've got a good drummer who probably won't bash your mics. If you see stick marks all over the place, watch out and pull the mics back.

One of the more famous minimalist techniques came from Glyn Johns, engineer and producer whose career started in the 60s working with major British artists such as The Who, Led Zeppelin, the Beatles, and so on. He accidentally discovered that one high mic facing down over the kit along with a second mic positioned just outside the floor tom, facing toward the snare, captured a huge sound from the entire kit. Both mics are roughly equidistant from the snare and panned accordingly to get the stereo spread you want. Quite a different concept than what we normally see today, but since it worked out pretty well for him maybe you should give it a go.

Piano

Condenser mics are ideal for piano due to its very complex sound and extensive frequency range. Place one over the high strings and another over the low end, but keep them a foot or more above the strings. The trick is getting a blended sound from the entire instrument. The lid can be a problem due to reflections, so often we'll remove it entirely, giving us more room to raise the mics higher. Another option is to stereo mic the piano from several feet away in the room with the lid all the way open. This results in a more classical sound. Leakage might be a problem if you're recording more than the piano at the same time. In this case, there are a couple of options. Lower the lid to half-stick and see how it sounds, or close it all the way and use PZM mics taped to the underside of the lid. PZMs are a very different sound, so it depends on the project.

Acoustic guitar

Even if the instrument has a pickup, don't use a direct box. Acoustics sound best when miked, so experiment with different condensers or ribbons about a foot or more away somewhere around the 12th fret (double dots). Too close to the sound hole and it'll sound very boomy, but too far in the other direction and you'll get mostly finger and string noise. There are other techniques engineers have discovered, but this will get you started. Compare a single mic with a stereo setup, such as an XY pair of large diaphragm condensers.

Electric guitar

There are two options for electric guitar—straight from the instrument into a direct box or placing a mic in front of a guitar amp. Often we'll use both, getting a clean sound from the DI while using the mic to capture the amplified sound of the cabinet (which is usually the preferred objective). So, what mics are good for guitar amps? The venerable SM57 has been a solid go-to mic for decades (because it's cheap and has a bump in the upper-middle frequency response that helps emphasize the presence and crunch sound of the amp). You can also find mics specifically designed for guitar amps, such as the Sennheiser e609. Place the mic in front of the amp grill an inch or two away, off-center of the speaker cone inside. Experiment by moving the mic closer or farther back and pointing toward various areas of the speaker cone. Try adding a second mic, either located the same distance from the cone or several feet farther back. Blending different microphone sounds and locations provides lots of tonal possibilities. By the way, listen to each speaker cone in the cabinet and make sure you don't mic one that's not working (more common than you think).

Personal confession—I once tried to put a very expensive condenser microphone on a guitar amp for a recording session. My logic was that I had this really nice, name-brand microphone (spelled N-E-U-M-A-N-N), and I figured it would blow away the cheap 57s that the band used. Boy did I get an earful...and yes, it was awful (I was very young and dumb in those days). Now, having said this, Alan Parsons (engineer for *Dark Side of the Moon*) likes condensers and ribbon mics on amps, so there's a reminder that there are no rules!

Bass guitar

For studio recording you have a choice of using a direct box and/or miking a bass cabinet. When running a live concert you want to avoid the cabinet at all costs, but unless there are other musicians in the room, miking a quality cabinet is desirable. The DI provides a clean sound that is easy to work with when mixing, but you need a quality model. Don't buy the cheap ones. For miking the cabinet, try a large diaphragm dynamic or condenser anywhere from a few inches to a few feet (or more) away. You'll get different sounds the farther the mic is located, particularly with bass frequencies.

Vocals

Most of the time a large diaphragm condenser is best for vocals, but you'll have to try different models to see what works best for a particular singer. Place it just above the mouth to avoid pops and sibilance, and a pop filter might be helpful as well. If the singer is using a music stand, try to angle it away from the mic to reduce reflections. Covering it with a towel can help also by absorbing high frequencies. While true for any instrument, pay special attention to the choice of mic preamp as it'll become part of the vocal sound.

Horns/woodwinds

Brass instruments are bright; sometimes we'll try a large diaphragm condenser, but often a ribbon works better due to their smooth, more mellow sound. For starters, position the mic just off axis of the bell a foot or two away. Woodwind instruments radiate sound differently due to the finger holes, so don't put a mic directly off the end of the bell. Try pointing it toward the lower third of the instrument, a foot or more away. This balances the breath, air, and bell sounds. As with anything, experiment with moving the mic a bit and you'll be surprised how much the sound changes.

Strings

These instruments sound scratchy up close, so give them some room. For violins, locate a condenser a couple feet over the player's shoulder facing downward toward the instrument. Cellos radiate outward from closer to the floor, so a large diaphragm condenser works well off to the side of the sound hole. Upright bass is similar to the cello, but play with the distance due to the longer wavelengths of low frequencies.

Small ensembles

This depends on how many and what instruments you're dealing with. We like condensers for their nice articulation and bright sound. They also pick up better from where you need to place them (several feet away), whereas dynamics don't do well at this distance. For an ensemble sound, try miking the group, not individual instruments; let the group create their balance, then you simply capture the magic. Use a stereo configuration several feet in front and adjust location to balance direct and room sound. If desired, add mics for each instrument in addition to the main stereo pair, balancing everything in the mix.

Backup vocals

These can be individually miked, stereo miked as a group, or even positioned in a circle with an omni smack dab in the center. Each provides a very different sound. Condensers, of course.

Overall ideas & suggestions

Depending on the situation, I often like to position portable diffusers around musicians. This creates a more controlled "room" that features a lively ambiance from the diffusion reflecting back into the mic. Stay away from windows and other flat surfaces.

The type of floor makes a big difference. Wood floors are great, and I've built drum platforms layered in 3/4 plywood that, as long as they're solid and don't give, will reinforce a solid low end. This works well for anything, not just drums.

No amount of miking technique and equipment will compensate for poor sounds in the room—it's got to be good at the source. Is it a quality instrument? Are the drums tuned correctly? Are the heads old? Are the guitar strings worn out?

Can the musician play well? I can't overstate that enough; you can hang a 57 on a string in the middle of a professional band and it'll sound amazing. I know from personal experience—the early sessions I engineered sounded pretty good, but it was because of the professionals in the other room.

Try different microphone preamplifiers as they all provide unique flavors. Also experiment with how hard you crank the preamp; some models sound great when overdriven while others can't take it.

Move the mics back and enjoy a larger sound. Too many times engineers close mic everything without even thinking there's a larger acoustic picture happening in the room. The early decades of British and American pop recording featured everyone in the room playing together, using the room acoustics as part of the sound.

Use the inverse square law to your advantage. In the near field, as you double the distance the level drops 6 dB. (This is the theory; it's actually 4dB or so in a real environment). This reduces leakage from other sources in the room. In the diffuse field (reverberant), this phenomenon doesn't apply.

Conversely, bringing a mic closer to a source increases gain to the console, reducing the amount of preamp gain required.

In terms of placement around the reverberation radius (boundary between direct and reverberant fields), directional mics can be placed farther into the diffuse field and still retain directionality. Omnis must be closer to the source since they are already picking up more diffuse field from the back.

Moving a cardioid mic even a little changes the frequency response and the instrument's sound.

Single-diaphragm omni mics possess directional characteristics from 1k —5kHz and up. At 16kHz they function basically like a cardioid. In other words, by turning up the high frequency on an omni you can make the mic more directional toward a sound source. (Dual membrane mics in omni mode don't have this characteristic).

Increasing high frequencies on an omni and a cardioid will bring out localization better—focusing that sound's placement in the stereo field.

Increasing bass frequencies will make the sound source seem closer. This won't work if the mic is in the diffuse field—you merely get more room sound.

Don't ever leave microphones on the floor—always in the box or on a stand.

THIRTEEN

STEREO MICROPHONE TECHNIQUE

Why mic something in stereo? That's how we hear the world around us. While we will talk a lot about recording ensembles and groups in stereo, using a stereo pair of mics on a single instrument provides a really nice sound that has depth and makes a big difference in a mix. Here's a quick example:

Audio example 81: Mono vs stereo miking an acoustic guitar

Perception of direction

How do we know where the car horn is coming from? Our two ears are designed to determine localization of sounds through two principles:

- Intensity difference
- Time of arrival difference

Intensity difference

Sound arriving from the right side reaches the right ear with a certain level of intensity (level). It must travel farther to reach the left ear, during which time the intensity has dropped a bit. This is due to the distance between the ears as well as physical blockage by the head. Also, a portion of the sound that reaches the opposite ear is from nearby boundary reflections, in which case the intensity level has dropped significantly. This occurs primarily in the mid and high frequency ranges, as low frequencies aren't affected nearly as much.

Coincident microphone techniques such as X-Y and Blumlein follow this principle. Both microphone capsules are located directly above each other, meaning they're not spaced apart horizontally. Therefore there is no difference in distance for sound to travel to either side. But, each mic is facing toward one direction of the source, thereby receiving those respective sounds directly. The other mic is facing off-axis to this direction, and therefore doesn't receive direct sound. This results in an intensity difference to the far-facing capsule.

Time of arrival difference

Sound arriving from the right side reaches the right ear earlier than to the left ear due to the spacing of the ears. Our brain calculates this timing difference and determines where the sound is originating. Near-coincident techniques (ORTF, NOS, OSS) follow this principle by slightly spacing the mics apart. (There is also an intensity difference due to the longer path length to the far side, so both principles are in play with these stereo configurations.)

Stereo imaging for mixdown

Following these two principles, the engineer can place sounds in the stereo field between two speakers, including in the middle where there is no speaker.

Panning

The panoramic potentiometer uses the intensity difference principle by proportioning more or less of a signal to one speaker as the engineer turns the pot. For example:

- If a signal is 5–10dB down in one channel, it will sound panned half-way to the other channel.
- If a signal is 20–30dB lower, it creates a full pan to the other side.

Delay

Using the time of arrival principle, delaying a signal in one channel will shift (pan) it to the other side. For example:

- A delay of 1/2msec pans it half-way toward one side.
- A delay of 1msec places sound completely to one side.

This holds true even if the other channel is up to 10 dB stronger.

Stereo microphone techniques: Coincident systems

The term *coincident* simply means the two microphone capsules are arranged over top of each other and are not spaced apart. Thus the diaphragms are arrayed in the same vertical plane with no difference in distance between the two mics. This means that directional cues come from intensity level differences only. The microphones are angled left and right to capture a stereo image. You can use two mics mounted together or one stereo microphone where the two capsules are built into the housing. The coincident technique provides more accurate L/R localization, but offers less sense of space and depth.

X-Y

The most common coincident arrangement features two identical cardioid mics with the capsules angled 90° (or wider). The capsules (not the body of the mics) must be positioned vertically over top of each other. If not, the slight distance between them will cause phasing issues. This setup will receive a left-right image from the front only, which eliminates much of the hall's ambient sound from behind.

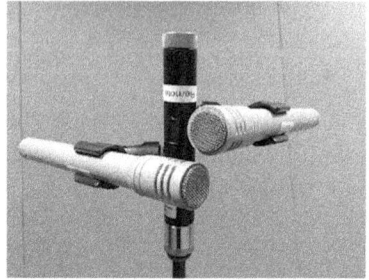

Blumlein

The Blumlein technique is simply an X-Y configuration with bi-directional mics. Capsules are angled at 90°, which receives a left-right image from the front and back; this results in a blend of the hall's natural reverberation which adds ambiance to the recording.

M-S (Middle-Side)

M-S features a single cardioid mic element facing the front with a bi-directional element facing sideways (left-right). A signal coming from the front left would be picked up by the cardioid and the front side of the bi-directional. These would add together and sound fine. However, a sound coming into the right (rear) of the bi-directional would tend to cancel when combined with the cardioid. (Bi-directional mics have a positive front and negative rear of the diaphragm.) Therefore the phase must be adjusted to allow both sides accurate pickup. Manipulating this phase and level balance between the capsules allows adjustment of the stereo spread, from mono to wide, without having to physically move the mics. M-S processing is very popular for mixing and mastering, and it was common for years in broadcast applications due to its rock solid mono compatibility.

If you don't have an M-S microphone, you can simulate the same effect with gear you already have in the pantry. Take a cardioid mic and face it forward, then place a bi-directional mic against it facing left/right. Find a way to split the signal coming from the bi-d mic, such as using a mult on the patchbay, so you have two copies of this signal. In a DAW simply duplicate this track. Bring them into two different channels, bus to two different outputs, and invert polarity on one channel. Pan them both center for now and bring both faders up until you get maximum cancellation—then mute. Bring the signal from the cardioid into a single channel on the console and bus it to the same two outputs as the bi-directional. Make sure levels are the same, unmute, pan the outputs. Of course you can use plugins to handle M-S processing as well.

Near-coincident systems

Near-coincident microphone arrays feature capsules that are not quite in the same space. The two mic capsules are placed a specified distance apart which allows time-of-arrival and intensity differences to help with stereo imaging. This method results in a better sense of spatial impression, providing excellent depth, but with somewhat less accurate L/R positioning. There are several variations of near-coincident configurations.

ORTF

This common near-coincident setup comes from the French National Broadcasting System. Two cardioid mic capsules are placed 17 centimeters apart (about 6.7 inches) with an angle of 110°. This spacing corresponds to the

distance between our ears. Low-mid frequency information comes largely from intensity differences, whereas high frequencies are determined through time-of-arrival cues due to their highly directional nature.

NOS

This variation on ORTF comes from the Dutch Broadcasting System. Two cardioid microphones are spaced 30 centimeters apart and angled at 90°. Low frequency phase differences occur lower in the frequency bandwidth. This method has a more open sound than ORTF.

Jecklin disk

This is an interesting derivation of the near-coincident technique where you mount a round disk between the two microphones. The disk must be rigid and covered with an absorptive foam layer. The reason for all this fuss? This method provides a very pronounced stereo separation which is quite effective. The photo here is looking down from over top of the Jecklin disk.

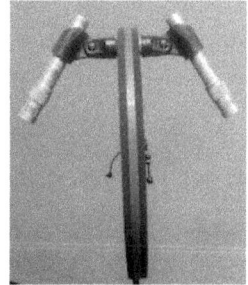

Spaced pair technique

Also known as A-B, this technique primarily follows the time of arrival principle, though intensity differences also contribute to localization. It is sometimes used in large rooms as the main stereo pair and is also very popular for room (ambient) mics. Smaller spacing of A-B is common for drum overheads. Spreading mics at large distances can introduce a hole in the middle; left-side information will go mostly into the left mic due to the inverse-square law, which tells us the signal will decrease before arriving in the right mic. You can place a third mic in the middle to fill in the hole, but keep the 3:1 principle in mind and don't place all three mics too closely together.

Other issues with spaced pair include a weaker center image, potential comb filter phasing between mics, and a phenomenon where a sound located on one side will audibly drift over to the other microphone. For example, a violin begins playing on the far left side. The A (left) mic picks it up and we hear it in the left speaker. However, as the sound continues it reaches the B-mic (right), though with a time delay. This causes the sound to appear to be moving across the stereo field from left to right, which of course the actual violin is not doing.

Decca Tree

One of the most effective ways to capture an ensemble, especially a larger, deeper group such as an orchestra, is with a Decca Tree. This technique was developed at Decca Records in the 1950s in order to capture an accurate stereo image with a strong center fill. It requires a special, T-shaped hardware positioner, with the base of the T facing straight toward the back of the source and the crossbar facing sideways. A minimum of three mics are employed, one placed center and slightly forward of the two side mics. The center mic is pointed downward toward the rear of the group, and the two side mics face down toward either side, angled outward about 45 degrees or so. This provides better coverage of the depth of the group while maintaining accurate stereo localization. One variation is to add two additional mics on the extreme ends of the tree facing back towards the audience for surround recording.

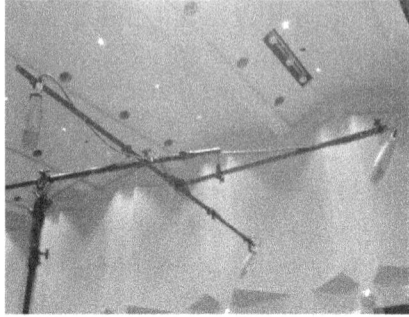

Binaural

Binaural recording is based on how the human head captures sound. In fact, these microphone systems have "ears" on either side with mic capsules inside, and may even feature an entire head assembly. It looks quite bizarre when mounted up front for recording something, but the results are fantastic. It's a very natural sound that's much more aligned with how we hear things around us. It's different from all other stereo techniques, so it's quite interesting to compare them. Binaural recordings are compatible with any stereo playback system, but the full effect is realized through headphones.

Recording with stereo mic techniques

So, what can you do with all this? Whether you are laying horn and string tracks for a pop album or recording a large symphony, understanding single and stereo microphone techniques will help you capture the natural, acoustic sound you're after. Your goal in recording acoustic sources is to hear everything in their proper location on stage with a clear frequency response. Sometimes this requires littering the stage with microphones, many times it can be done with just two mics located in the right spot. Learn the fundamental rules, such as 3:1 spacing, phasing issues, and room acoustics, then use your ears to determine what works and what doesn't for any given situation. First we'll provide a brief overview of how sound propagates in a room that will help dictate where to place microphones. Then we'll discuss some general guidelines for recording large ensembles, smaller chamber groups, and even single instruments.

Basic room acoustics

Sound generated in a space will spread out and reflect around the various surfaces. Size, shape, and surface treatments determine frequency response and the nature of these reflections, meaning they differentiate a bathroom from a concert hall.

A propagated sound passes through three stages:

```
┌────────────────┐      ┌────────────────────┐      ┌──────────────────┐
│ Direct sound   │ ───▶ │ Initial reflections│ ───▶ │ Reverberation    │
└────────────────┘      └────────────────────┘      └──────────────────┘
```

Direct sound is what you hear straight from the source, with no reflections from nearby surfaces affecting it. The closer you are to a sound source, the more direct sound you will hear.

Initial reflections are the first waves to be reflected from nearby surfaces, primarily the front and side walls. The delay time between the original direct sound and these early reflections provides us with a sense of how large the room is. Think about it—in a larger room the walls are farther away, so it takes longer for a reflection to bounce off the wall and arrive at your seat. Therefore a longer initial delay sounds like a bigger room.

The reflections in a room rapidly multiply, particularly when the source continues to produce sound. As the reflections become more numerous and dense the result is *reverberation*. This evenly distributed, diffuse sound field contains no discernible reflections and will eventually die away at a rate dependent upon the physical attributes of the room (size, wall surface textures, and so on).

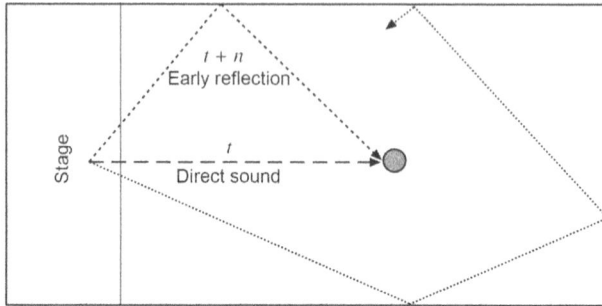

There's an acoustic dividing line between direct and diffuse sound, and you can actually hear it. Walk around a performance hall while an orchestra or choir is rehearsing. Walk slowly toward and away from the stage. Eventually you should be able to find a spot (roughly 1/4 or 1/3 of the way into the room from the stage area) where you can hear mostly reverberation if you lean back, but more direct sound from the stage when you lean forward. This boundary is termed the *reverberation radius* or *critical distance*, and it turns out to be the ideal spot to locate your main stereo mic pair for an ensemble. Start here with your setup and then listen. If it sounds too close, move it back. If too reverberant, simply move it closer. This takes a lot of the guesswork away. If you are using omni-directional mics you'll want to locate them a little closer to the stage since they'll pick up more room reverberation from the back.

Another method for finding the critical distance is to put up two mics— one in the diffuse field, the other a little closer to the stage. Slowly move the second microphone closer to the stage while watching your meters; when you see a relative difference of 3dB between both mics, you've found the boundary. Use the same microphones that you plan to record with.

Typical large-scale hall setup

If you're recording a fairly large source, such as an orchestra, concert band, or choir, follow these guidelines for selecting and positioning microphones so as to capture the entire group as faithfully as possible.

Main stereo pair

Choose between a coincident, near-coincident, or Decca Tree approach. All mics must be identical condenser-type models. This main stereo configuration should be located near the ensemble within the critical distance boundary, not back in the hall as many novices tend to do. Place it relatively high, somewhat over the conductor's head, pointed down toward the middle or so of the ensemble. Move it closer or farther away to achieve the balance of direct and reverberant sound you want.

Support mics

When miking a larger ensemble, it's easy for quieter sections or soloists to be overshadowed. Engineers can place a microphone close to that source and blend it into the overall mix. This is referred to as *support miking*, and the trick is to get them to blend in acoustically with the main stereo pair. Use a large-diaphragm directional mic and close-mic the instrument. Pan the signal true to the overall stereo image (as captured by the main stereo pair), and run it through a digital delay to compensate for the difference in time of arrival to the support mic and the stereo pair (1ms per foot of distance). This keeps it from sticking out, maintaining its place within the ensemble.

Room mics

Consider placing a stereo pair far back in the hall to capture ambiance in the reverberant field. Typically a spaced A-B is used, though two boundary mics on the back wall work fine. The quality of the hall's acoustics will determine whether these are used or not. When using spaced A-B there are a couple things to keep in mind. Keep them several feet away from the walls (unless you're using PZMs), and less than 1/2 distance vertically between the floor and ceiling (to avoid standing wave nodes).

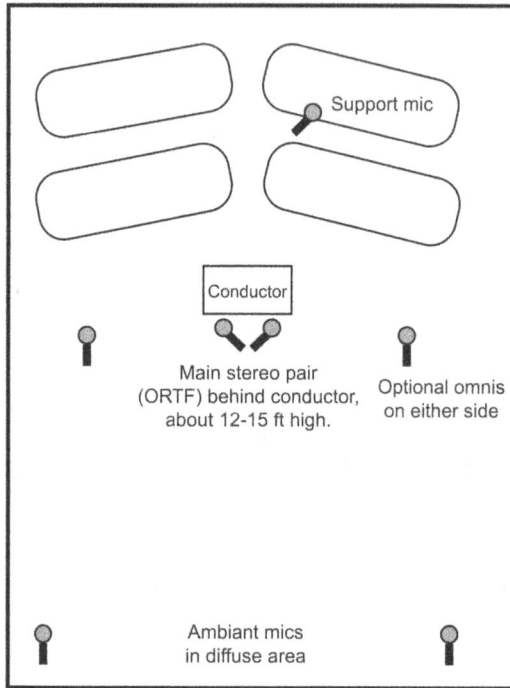

With all of these mics, experiment with spacing and distances so as to capture an accurate, clean stereo image of the ensemble. If the instruments sound too scratchy or close, raise the mics a little. If you're getting too much hall reverb then you're too far away. A small change can make a huge difference. Some engineers might also add two omni-directional microphones on either side of the main stereo pair, closer to the ends of the ensemble. These provide additional depth and breadth from the group while adding some natural reverberation from the hall.

Typical chamber group setup

As usual, begin with a primary stereo pair up front, not too far from the group. Actual height and distance depends on the nature of the room. Two additional omnis can be placed out to both sides to add depth and ambiance. It's possible to individually mic each instrument, but this becomes more complicated for trying to get a natural mix from the group. By simply using a single stereo pair you let the ensemble control their own balance, which they should be able to do if they're good players. This also allows the total sound from the group to blend and mesh as it travels toward the mics, which provides a more natural sound. We should also mention that it helps to record a chamber-style performance in a smaller room, thus mimicking the type of space these groups would use. The term chamber music comes from the days when small ensembles, such as string quartets, would perform in living rooms. This is a very different sound than that of an orchestra blasting the 1812 throughout a large concert hall.

Listen for the blend between direct and diffuse (room) sound. If it sounds too close and in your face, move your mics back and/or higher. If it sounds distant and muddy, you probably want to move the mics closer. Chamber music is best reproduced fairly close, mimicking the environment of a small room.

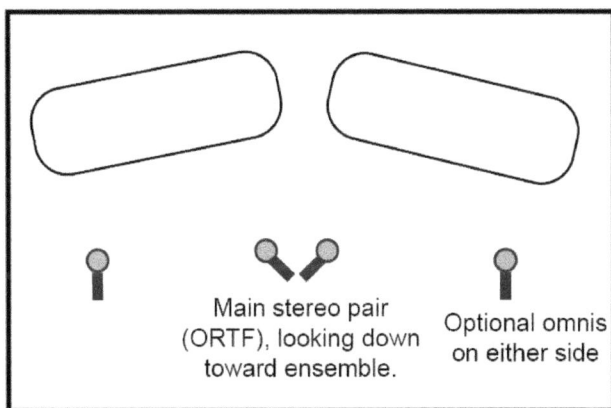

Main stereo pair (ORTF), looking down toward ensemble.

Optional omnis on either side

In the studio

Stereo miking is not just for concert halls; it's a great way to capture small groups in the studio. For example, use a stereo setup to record a horn line or backup vocal section. Drum sets usually involve a stereo pair overhead or out in front of the kit.

Putting a stereo miking setup on a single instrument may sound odd, but it actually sounds really nice. Try comparing a single mic on an acoustic guitar, for example, with a stereo configuration and you'll hear it immediately. Two condensers set in XY or ORTF provide more spatial dimension. You probably don't want to pan these hard left and right, though, so use your ears to see how to fit it into the mix.

FOURTEEN

SIGNAL PROCESSORS—FREQUENCY

We can examine sound in three dimensions: frequency, amplitude, and time. Each of these can be manipulated with signal processing, affecting a sound's timbre, volume envelope, and how it develops over time before dying away. The next three chapters explain the relevant processing for each domain and how they are applied in the studio.

Equalizer

An equalizer, or *EQ*, is a circuit that changes the frequency response of a sound by boosting or attenuating selected frequency bands. Think of it as a sophisticated tone control that allows you to make a sound brighter, less boomy, and so on. To understand how it works, you need to grasp the concept of what sounds are made of.

There is a wide range of frequencies that are audible to humans—anything vibrating between 20Hz and 20kHz falls within our hearing range. All sounds coming from musical instruments, voices, or traffic on the highway produce a number of frequencies which fall somewhere within that band. What makes a trumpet sound different from a fog horn is the difference in *which* specific frequencies and *how much* of each frequency is included in that sound. In other words, the harmonic content of that sound is what makes it unique. We call this the *timbre* of a sound.

With an EQ, we can actually raise or lower these frequencies within a sound—this alters the harmonic structure and therefore makes it sound different. You're not turning a flute into a guitar, but you can make the flute brighter by boosting that sound's upper frequencies.

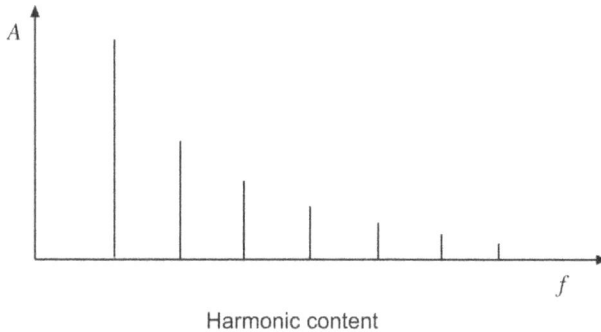

Harmonic content

Equalizer types

There are a few main EQ designs that give you control over a signal's harmonic structure, and each of these employs different types of filters, which are the circuits that actually do the work. First, let's introduce the EQs you'll run into, then I'll explain the filters they use.

Bass & treble

The simplest EQ around, this type gives you control over the treble (high frequencies) and the bass (low frequencies). That's it. This is the most common type found on home and car audio systems. They use shelving filters (explained a bit later) and function similar to the high and low frequency controls on your console or plugin. They're based on the Baxandall EQ curve, developed in 1950 by Peter Baxandall in England, which features a very gradual boost or attenuation.

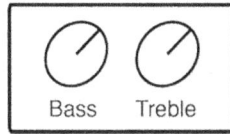

Bass   Treble

Graphic

Graphic EQs are easily identified by the row of vertical sliders that are used to boost and attenuate specific frequencies. These sliders are preset at certain intervals (frequencies) throughout the frequency spectrum, so you can only select what's there to change its level—you can't change the exact frequencies. These are found in some home audio systems, car stereos, as well as professional applications. They're sometimes used for "tuning a room" when an engineer adjusts the speaker system to compensate for acoustic issues in a particular room or performance hall. You determine which frequencies are causing problems, such as persistent feedback, then reach for that particular slider to turn it down. Most digital consoles provide graphic EQs on buses (aux sends, groups, main mix).

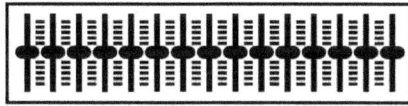

Graphic EQ

Parametric

Parametric EQs are the most common type found on mixing consoles. There are parametric EQs for each channel, though with digital consoles you'll typically have them available on all channels, sub-group buses, aux buses, and the main mix. Of course with DAWs they're available as plugins to insert wherever you like. This is the most complex type of EQ, usually employing both peaking and shelving filters, and provides not only the option to boost or cut certain frequency ranges, but also to dial in the exact frequency region you want to work on. Whereas graphic EQs come preset for certain bands, parametrics allow you to move around and find where you want to work.

Parametric EQs divide the audio spectrum into a number of bands (regions). You always have high and low frequency controls; sometimes these are fixed at a specific frequency, other times they are adjustable. At least one mid-range band will provide boost/cut along with frequency select, allowing you to fine-tune exactly which frequency area to work with. Often you'll get a third control, bandwidth, that determines how wide an area to affect.

On the diagram below, look at the section labeled "low-mid freq" and you'll see these three controls. On consoles, usually the mid-range EQ uses peaking filters while the high/low controls are shelving. Sometimes this is switchable between shelving and peaking—check the console or the manual.

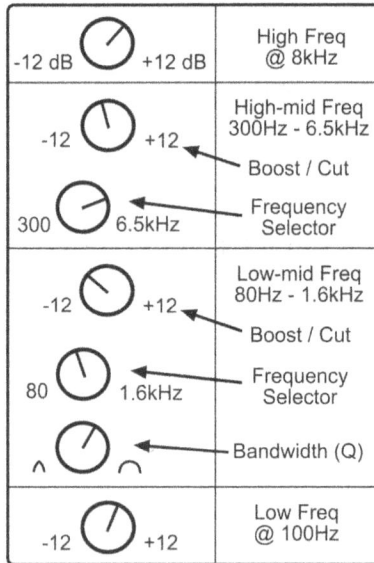

Parametric EQ Module

Parametrics give you very precise control over your sound, but it takes some understanding and practice to use them well. Play some individual tracks and experiment with the various controls. Keep reading through this section and I'll show you more of how they work and what they sound like.

Active dynamic control

There are two different concepts to distinguish here. Lots of different audio devices, including direct boxes, speakers, and processors, come in either passive or active designs. Passive models employ capacitors, inductors, and resistors to do their work, whereas active devices have transistors or tubes that provide amplification and control of electrical signals. There's nothing inherently negative with either approach as they feature different sounds and results. Engineers have been using active and passive EQs in the studio for decades. For example, the passive Pultec EQP-1A has long been famous for its huge bottom end, ultra-smooth response, and musical sound. (It has a tube on the output for amplification, but this is for color and make-up gain, not part of the filter circuits.) EQs with active components use transistors or tubes in the filter circuits, which color the sound a bit. But, the overall concept for both of these designs is the same—you're merely setting a fixed adjustment of frequency component amplitude in a sound.

Now, what we're talking about here are software EQs that feature active dynamic control over the EQ processing, which is a fairly recent development. With a standard EQ, it will always boost or attenuate a certain frequency range to a set amount no matter what's happening with the audio signal. Wherever you set the slider or knob is how it will continuously affect the sound. An active EQ functions much like a combination EQ and compressor. You set the frequency regions where you want EQ to operate, but these adjustments only kick in when an amplitude threshold is met. This is a much more intelligent and musical design in that it makes a difference only when needed.

Linear phase

Traditional EQ circuits introduce a certain amount of phase shift within the audio signal. As part of the filtering process, tiny timing adjustments occur as it attempts to rebalance frequency components within a region. Of course changes in timing affect the overall coherence of the signal (phase), so it further alters the timbre in a way that sometimes doesn't sound that great. Engineers have lived with this forever, so it's just part of life; there were times when I decided against using an EQ at all because the cure was worse than the problem I was trying to fix. On the other hand, this is partly how an EQ adds color to a signal, so that's a good thing. Design engineers have worked to minimize this issue, resulting in what they generally term *minimum phase EQ*. This includes most of the EQs you'll find in your rack or plugin collection.

Linear phase EQs eliminate phase issues by approaching things in a completely different way. They delay the signal so that changes can be processed, taking time to crunch the numbers, then output the result in a time-coherent signal. Processing is completely transparent with no coloring of the sound; it's generally used for fixing problems, such as surgically removing a narrow band. Is it better than traditional EQ? As always, it depends on the situation. Most of the time you'll want to use normal EQs on your tracks due to the sound they provide. Linear phase models are terrific for targeting problems; other than the lack of the traditional EQ sound, the main side effect is a fairly significant processing delay (latency), which prevents them from working well in real-time situations.

Filters

Peaking filters

Peaking filters allow you to boost or attenuate a range of frequency components centered around a specific point, called the *center frequency*. You can't operate on just one frequency, though—it always affects a certain number of frequencies on either side. This creates a bell-shaped curve and is known as bandwidth. The bandwidth of frequencies affected is called Q, for quality factor, and it can be widened or narrowed in many parametric EQs.

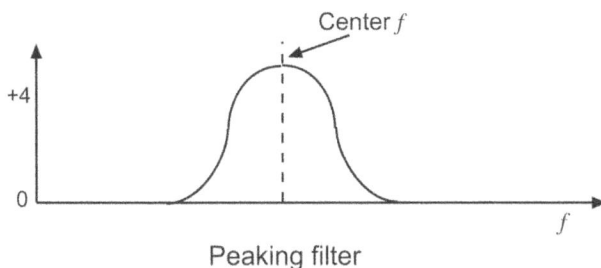

Peaking filter

The term Q is easier to use instead of precise frequency counts because the number of frequency components per musical octave doubles as you go up the scale. So, if you boost an octave in the low end, you might be adjusting forty frequency components, whereas in the high end you could be working on eight thousand frequencies.

They both sound like musical octaves to us, so we need a system that compensates for the difference in actual frequency components. Q does this quite nicely by providing a simple number to refer to; a Q value of 0.8 sounds exactly the same regardless of where in the frequency spectrum you're working. Most consoles will label this control either with the numeric Q value or with a graphic that looks like a bell-curve.

All this means is that you can dial into any specific frequency location you want to fix or enhance. It might be a small slice, such as reducing a 60Hz hum, or it might be a wide chunk of your sound, such as boosting high frequency sparkle on the piano.

Use the following formula to calculate Q:

$$Q = \text{Center frequency/Bandwidth}$$

Example: If the center frequency is 1kHz and the filter is 1/3 octave wide, and its bandwidth lies between 2.5kHz and 3150Hz, then the bandwidth = 650Hz. Therefore, divide 1000Hz by 650Hz, resulting in a Q of 1.54.

- Larger Q values = narrow bandwidth
- Smaller Q values = wide bandwidth (affects larger range)

A couple more examples:

$$Center\ frequency = 1kHz$$
$$Bandwidth = 10Hz$$
$$1000\ /\ 10 = 100\ (large\ value)$$

$$Center\ frequency = 1000\ (1kHz)$$
$$Bandwidth = 3000\ (3kHz)$$
$$1000\ /\ 3000 = .333\ (small\ value)$$

Shelving filter

Shelving filters are often used for high and low frequency controls. They provide a boost or cut at a selected frequency (*turnover frequency*) at which point the adjustment remains constant throughout the rest of the spectrum. So, while a peaking filter will affect a range of frequencies and taper off on either side of your selected band, the shelving filter affects everything equally from the turnover point and beyond.

On the diagram below, the vertical axis represents amplitude (signal level) and the horizontal value is frequency range from low to high. Everything to the right of the turnover point is boosted. Everything to the left of this remains at 0 (unity), meaning the original level passes through unchanged. Note that the actual turnover frequency is 3dB down from maximum boost or attenuation, with the shelf flattening out beyond that point.

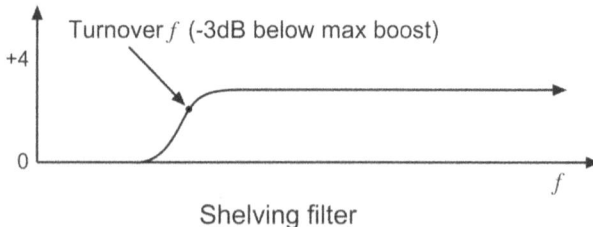

Shelving filter

Notch filter

Notch filters are peaking filters with an extremely narrow Q. They are designed to cut deep into a sound to eliminate a specific, narrow frequency range. One common application is getting rid of a 60Hz hum from a ground loop. I've also used these to notch out ringing overtones from cymbals and even guitar amps. The idea is to surgically remove an offending sound without interfering with the music if it can be helped.

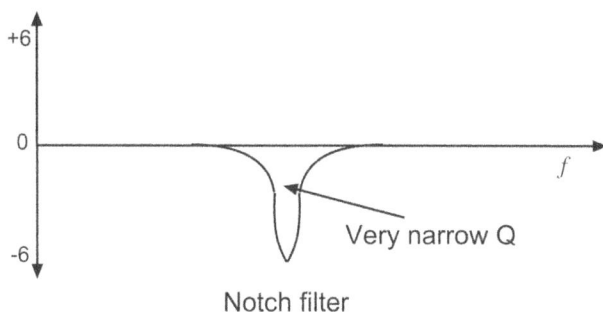

Notch filter

High- & low-pass filter

These work differently from the others in that they provide no boost function. All they do is sharply attenuate frequencies above or below a certain point in the frequency spectrum. A high-pass filter set at 100Hz (*cut-off frequency*) will attenuate all frequencies below 100Hz, and a low-pass filter set at 8kHz will attenuate everything above this point. The rate of attenuation beyond the cut-off frequency is called the *slope*, which can vary between 6dB/oct, 12dB/oct, 18dB/oct, or 24dB/oct. Thus for every octave beyond the cut-off frequency the signal drops 6, 12, 18, or 24dB; higher numbers result in a quicker attenuation. There is no way to design a perfect "brickwall filter" that magically slices everything at the cut-off frequency. Actually, the cut-off frequency is about 3dB down from unity gain, meaning the attenuation begins slightly before this frequency.

Low-cut filters (also called high-pass) are great for getting rid of room rumble, low-frequency leakage from other sounds in the tracking room, ground loop hum, and that annoying pop from a vocal getting too close to the mic. All of this noise interferes with the music waveform, which your speakers are trying to reproduce. Consoles provide a low-cut filter on every channel, with the cut-off frequency usually preset somewhere around 80Hz on analog boards. Digital consoles and plugins allow you to dial in any cut-off frequency you want. Most of your sound sources do not have anything below 100Hz, so you won't miss anything by getting rid of it. In fact, I use these on nearly all my tracks when mixing, the exceptions being bass guitar, kick drum, and so on. The result is a cleaner sound without the low-end mud.

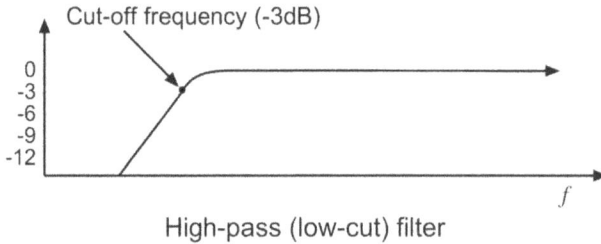

Cut-off frequency (-3dB)

0
-3
-6
-9
-12

f

High-pass (low-cut) filter

Same, but different

Manufacturers have developed different approaches to the task of EQ for a reason—each one brings a unique flavor, advantage, and result to the table. Just because two models feature similar-looking parametric peaking filters in the mids doesn't mean they sound the same. Take a look at the following two diagrams:

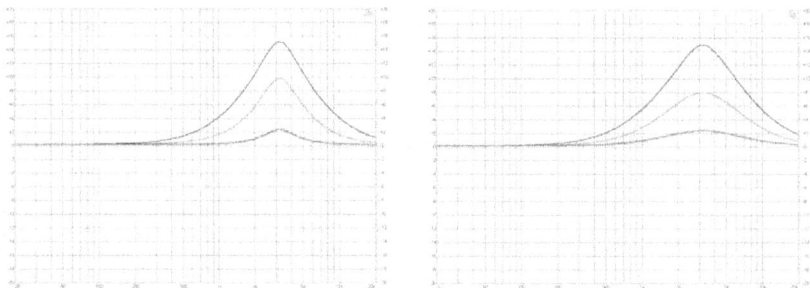

The left diagram features sharper curves and peaks, focusing closer around a particular center frequency. The EQ on the right provides a gentler, smoother curve and is generally considered more musical. Both are excellent designs, but they operate a bit differently and give you very different results. Once you understand this concept, use your ears to select which type works for any particular situation.

Application

EQs are used to address tonal problems such as mic placement coloration or the physical setup of the instrument. We can also creatively change the sound of an instrument to fit our mix better. If there is a muddy region in the piano mic, and moving the mic doesn't help, then attenuating frequencies at that region can clean up the sound. For more presence and sparkle from the piano, boost a little in the high frequencies with a shelving filter. If you need to make a vocalist sound fuller with more bass, then boosting lows can help to a certain extent. But, if there aren't any low frequencies, then it does no good to try boosting something that doesn't exist. I once had an artist who kept demanding more bass on the vocal—and she was a soprano!

EQ should not be expected to compensate for poor quality sound sources or bad miking technique. Don't automatically reach for the EQ —go in the tracking room and listen to the instrument. How does it sound in the room? Might it sound better in another location? Many times it's the instrument itself that needs work, so collaborate with the musician to see what they can do. Fix the instrument, find the perfect location, then pick the right mic and play around with where to place it.

Many engineers won't EQ during a tracking session. The idea is that you don't yet really know how it will blend with the final collection of tracks. I've gradually become more of a minimalist during tracking, preferring to get my sounds from the room, the mic, and the mic preamp, but it's not wrong either way.

It's also interesting to see how different engineers use an EQ. Some prefer as little as possible, not wanting to interfere with the sound they have. There's a great Elton John record that was mixed with no EQ (no other processing, as a matter of fact) other than a tiny bit of high frequency boost on the drum overheads. It always goes back to your raw sounds, musicians, and song arrangements. Then there are other engineers who crank the knobs as far as they can go. Whatever floats your boat, but the difference is that professionals understand what they're doing and how it affects things. Learn the underlying concepts, develop your listening skills, then trust your ears.

A few remaining points to remember

Many engineers try to attenuate first; humans don't hear a cut as easily as a boost, so this is a more transparent approach to altering a sound. For example, you can brighten a track by attenuating low-mid frequencies. This can clean up the overall sound, allowing the upper-mids to shine through with greater presence.

Boosting frequency regions with EQ adds to the signal's overall dynamic range and level. Go too far and it will distort the audio circuits (plugin, console, etc). Always keep an eye on your meters.

High-pass filters reduce low-frequency leakage and noise, dramatically cleaning up your entire mix.

Some EQs are better at focusing on specific problems, others more for tonal shaping. I often use two on a track, the first to fix any issues and the second later in the processing chain to provide musical color in the mix.

Different EQ models bring a unique flavor to your tracks, so experiment.

Let's hear what they sound like

These first examples have settings that are fairly extreme in order to make them easier to hear.

Audio example 82: Flat EQ

Audio example 83: 9dB boost at 1kHz

Audio example 84: 9dB attenuation at 1kHz

Audio example 85: 9dB boost at 4kHz

Audio example 86: 9dB attenuation at 4kHz

Audio example 87: 9dB boost at 8kHz

Audio example 88: 9dB attenuation at 8kHz

Audio example 89: 9dB boost @ 125Hz

Audio example 90: 9dB atten @ 125Hz

If your console provides bandwidth control, you can choose to adjust a big chunk of the sound or just a small slice. The three examples in this file feature no EQ, EQ cut with a pretty wide bandwidth, and then EQ cut with a much narrower bandwidth.

Audio example 91: Low-mid attenuation—flat, wide Q, narrow Q

Here's the difference between a high frequency peaking and shelving EQ. The peaking filter has a slight boost in the mid-range—it makes the guitar a little edgier. Then we switch to shelving, maintaining the same 3kHz frequency point. The difference is that it boosts everything above this point, so you hear all the noise and other nasties way up there. In this case we would prefer the peaking EQ, but it depends on the situation. Drum overheads, for example, usually work well with shelving, as do vocals, pianos, and so on.

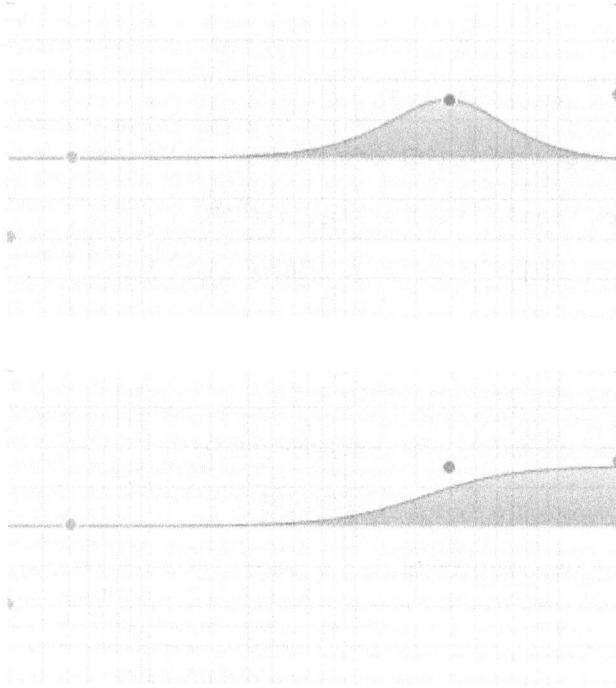

Audio example 92: Hi-freq peaking, hi-freq shelving

How to set an EQ

Some EQ adjustments are pretty straightforward. Need to make a cymbal or tambourine brighter? Turn up the high frequency control. For some consoles the high and low ranges only have boost/cut controls at a fixed frequency, such as 8k or 100Hz.

For the mids, which generally have more controls to work with, many engineers find that it's easiest to over-boost EQ in the range they're targeting, focus on the exact frequency area that needs work, then make the final boost or cut adjustment. If a vocal needs a bit more presence, turn up the high-mid frequency gain control several dB so you can clearly hear the difference. Now turn the frequency select control back and forth to move through the frequencies. Listen for a spot that sounds good for what you need, then pull back the boost to a good level.

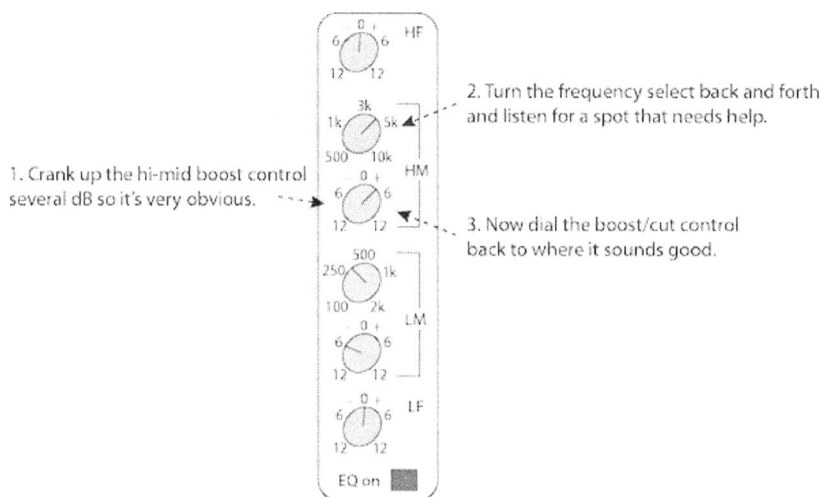

1. Crank up the hi-mid boost control several dB so it's very obvious.

2. Turn the frequency select back and forth and listen for a spot that needs help.

3. Now dial the boost/cut control back to where it sounds good.

For this next audio example, we've looped a guitar part four times. First we boost low-mid EQ a lot so we can hear it, then sweep up and down looking for a muddy section. Hear how it gets really boomy and muddy near the end of this first loop? We pull it down below zero to attenuate it 3 or 4dB. The second loop starts bypassed so you hear the original sound, then we cut in the low-mid EQ band around the halfway point. Now for the third loop we do the same for the hi-mid band, looking for some natural, nice presence and detail. I like the sound of the picking and strings just under 4kHz, so we bring down the boost so it's up about 2 or 3dB. The last loop starts completely bypassed, then we kick in the entire EQ. It might sound a bit thin and bright by itself, but depending on the song it should fit in nicely and cut through the mix.

Audio example 93: EQing an acoustic guitar

Reduce muddiness

Add presence "brightness"

HF
3k
HM
LM
LF
EQ on

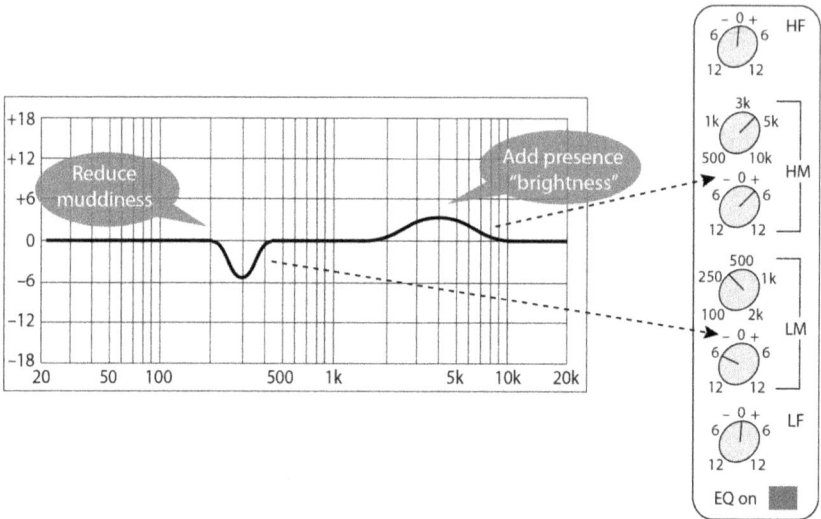

This process takes a lot of practice to hear what's going on and understand how the controls affect it. Over time you'll be able to hear what's "normal" and what's "not what you want".

Here's an entire drum set without, then with EQ on each mic channel.

Audio example 94: Entire drum set before/after EQ

Reducing a muddy low-mid range on a vocal to clean it up while adding some hi-mids for presence.

Audio example 95: Dipping "mud" from vocal (flat, EQ, then with compression)

Audio example 96: Same thing for a female vocal (flat, EQ)

And a few more...

Audio example 97: Bass guitar (flat, EQ, EQ/compression)

Audio example 98: Acoustic guitar (flat, EQ, EQ/compression)

Audio example 99: Electric guitar (flat, EQ, EQ/compression)

Key frequencies for various instruments

- Kick: body around 80Hz, attack around 2k
- Snare: body 200–300, attack and crispness around 2–4k
- Cymbals: high end shimmer 7–12k
- Mounted toms: body around 200–300, attack around 4k
- Floor toms: body around 100Hz, attack around 4k
- Vocals: intelligibility 2–4k, presence 4–5k, sibilance 6k
- In general, many instruments have a body sound in the 200–300 range, with attack and/or articulation transients in the mid-freq range (2–5k).
- Muddy sound generally happens somewhere between 200–400Hz, give or take depending on the instrument. We often pull this area down to clean up the sound and make it sound less cloudy.
- 500Hz often features a nasal, honky sound that should be removed with a narrow-band attenuation.

Enhancer (aural exciter)

Aural exciters work differently from EQs in that they don't boost or cut existing frequency components. Instead they actually generate extra harmonic components based on what's already there. They're great for making a sound brighter and edgier by adding upper harmonics. If you have a track that sounds kinda dull, or even an entire mix, throw an exciter on it.

---

*Audio example 100: Adding varying amounts of aural exciter to
a track*

Harmonizer / pitch shift / pitch correction

Harmonizers have been around a long time. They analyze the funda-mental frequency (the primary pitch) of the sound and can then either detune it (to fix an intonation problem) or generate harmonies on top or below. Once when I was really young I tried using a harmonizer to match a piano track to the band. Didn't work, of course; they need a single pitch to lock in on. Much later in the 90s I remember visiting a friend in Nashville and poking my head in a back room. I guess it was an intern sitting there staring at a computer screen, every so often clicking once or twice, then staring some more. He was adjusting pitch for each vocal note of the album, one at a time. Excruciating.

Antares Auto Tune is the most recognized product for DSP-based frequency analysis and correction. The original intent was to adjust minor pitch anomalies for vocal recording, but it was soon discovered that if you pushed it aggressively the result was a robotic sound. Even with normal operation it can leave artifacts behind. So, the better your musician is, the less correction is needed, and you get a cleaner sound.

Tone & transient shaping / tape emulation

Modern DSP has brought lots of variations on the concept of frequency processing, allowing us to dynamically manipulate pitch, transients, and other parameters to give a track more life, punch, presence, and so on. These can blur the boundaries between traditional signal processing, combining functions of EQ, compression, delay, harmonic generators, etc. Tape emulators attempt to model characteristics of analog machines, many of which were largely considered limitations: tape noise, harmonic distortion, bias calibration, tape speed inconsistencies, and many more. It's a complicated combination of factors that adds up to that desired "analog warmth", and frankly they work pretty well. And far cheaper than maintaining and operating a vintage tape machine.

Audio example 101: Tape saturation for warmth

FIFTEEN

SIGNAL PROCESSORS—AMPLITUDE

The second of the three domains we can examine in an audio signal is *amplitude*. This is the difference between a zero reference point and maximum electrical voltage or displacement of a vibrating object. A guitar string just before it's plucked is at a state of equilibrium, or zero amplitude. When plucked, the string extends beyond this point, and the harder it's pulled the farther it vibrates—making it louder to our ears. Electrical signals going through a console or other audio components are the same thing, where amplitude is based on voltage level.

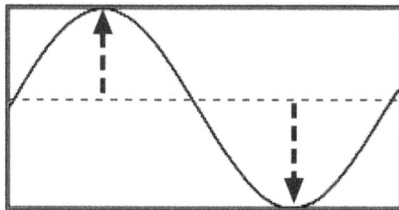

All components of an audio system, including the microphone, console, recorder, monitor speakers, and even your hearing, have varying capability for handling signal amplitude from the lowest level to highest peak. This is referred to as *dynamic range* and is indicated as a *signal-to-*

noise ratio (S/N). The level of a signal is measured between the noise floor of that particular component and the maximum level achieved just before the system goes into distortion. Modern equipment provides sufficient dynamic range as long as proper levels are maintained throughout the signal chain; it is the recording medium that has typically been the primary issue over the years. Professional analog tape recorders achieve an S/N of around 60–70dB, whereas the compact disc extends this to 100dB; modern 24-bit systems far surpass even this at 144dB S/N. An orchestra performance might possess a dynamic range of 65dB, but during low passages the noise floor of analog tape becomes noticeable. There is no noise floor in digital recording of any consequence, so the extended dynamic range allows us to record a wide-ranging performance without fear of distortion or noise intrusion (assuming recording levels are carefully set and monitored).

There are times when control over a signal's dynamic range becomes necessary. Radio requires heavy limiting due to restricted broadcast dynamic range. Noise floors for analog tape, and especially the consumer compact cassette, are fairly high and thus require some constraints on a signal's dynamic range. However, these days the issue is not so much needing to control amplitude ranges due to equipment limitations as it is fitting everything into a mix while seeking musical and sonic creativity.

A musician's performance usually varies quite a bit between soft and loud notes. Even during a relatively consistent rhythm part, such as a strumming guitar, performance fluctuations can cause significant differences in level. The same goes for vocalists, who may sing the verses normally, then kick into high gear on that dramatic final chorus. The result is that instruments and vocals will jump in and out of the mix, so some control over dynamic range will help keep things in their place.

Sonically, altering the dynamic range of a signal provides interesting timbral results that add to an engineer's options for crafting a mix. Because of this, many different types of amplitude-based processors have been developed over the years. Changes in electrical components, circuit design, and other factors are what give each model its special flavor. You should invest sufficient time trying a variety of processors on a track to get a feel for how they sound. Sometimes engineers aren't even looking for gain reduction, but to just color the sound in a pleasing way. Let's take a look at what a *compressor* does and how to use it, then listen to what it can do for your music.

Compressor

A compressor is a signal processor that limits how much an audio signal varies from soft to loud. As the incoming signal gets higher, the compressor will reduce this so it sounds more even. Think of it as a volume cruise control for audio.

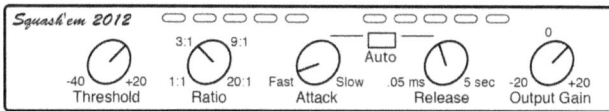

Compressor functions

Threshold is the signal level at which the device begins to compress. Any signal that is higher than the threshold will get attenuated, meaning it won't be so loud coming out the other end. Threshold doesn't directly dictate how much compression will be applied, only the point at which the device starts doing its thing. A high threshold will only compress the highest, loudest levels in the signal, whereas a lower threshold affects more of the entire signal.

Ratio sets how much the signal will be reduced. Once a signal goes over the threshold point, it will be attenuated at a rate set by the ratio. For example, if you set it at 3:1, every 3dB of signal beyond threshold will result in only 1dB coming out of the compressor (smaller dB numbers are quieter). Since these are ratios, multiples apply here, meaning if 9dB of signal goes over threshold, it will output 3dB.

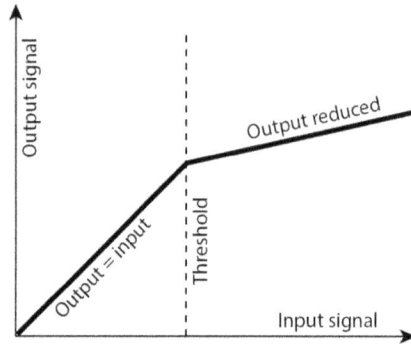

A signal going over threshold triggers the processor to start compressing, but we can control how quickly it gets to its full reduction setting. This is the *attack* time; a fast attack setting will very quickly reach maximum gain reduction once threshold is reached, whereas a slower attack will take its own sweet time to get there. This is a big deal because it affects the tone of the sound as well—a really fast attack will lose the initial bite or attack of the sound, giving you a more rounded tone. Lengthening the attack time will allow that bite to come through before the overall dynamic range gets limited. Many people misunderstand this concept, believing that a longer attack time means the processor sets a timer before beginning the compression cycle. Not so. The technical term for what you're shaping is called *transients*, which are the very first high frequency components of a sound that happen when an instrument is plucked or hit. So, turning the attack time up and down will give you a brighter or duller tone.

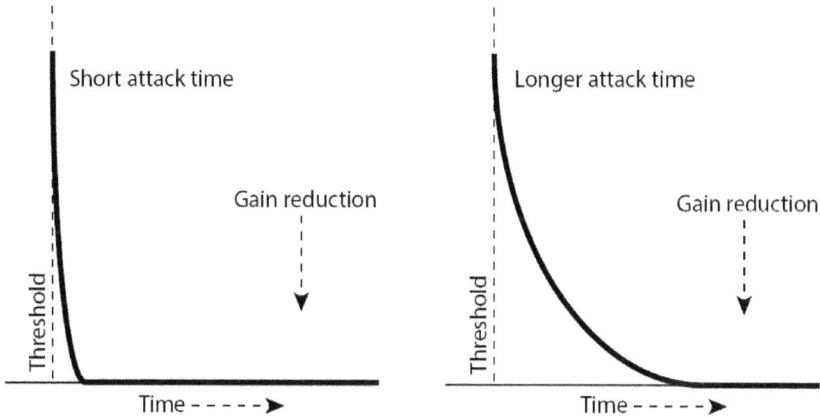

Many compressors also provide a *knee* setting, which effectively broadens the threshold point into more of a range. Once a signal approaches threshold, the processor will initiate a smooth, gradual transition into the compression cycle.

Once the incoming signal begins dying away and falls below threshold, the compressor will "let go" and allow the signal level to return to normal (unity gain). How long does this take? That's the *release* time you set. A medium release is usually pretty smooth, whereas a long release may prevent the unit from recovering before the next musical passage begins. This obviously depends on the tempo and what they're playing. A long release could also extend the life of a note just a bit; as the original sound fades away naturally, the compressor is slowly returning to unity gain, briefly counteracting the decay. A very short release setting can cause the unit to cycle too quickly, resulting in a breathing or pumping sound. Most compressors feature an auto setting option for attack and release, and for the most part it works pretty well. However, go ahead and take control yourself and experiment to make it fit each particular situation.

After a signal goes through a compressor, it's usually got a lower overall signal level and might sound quieter in your mix. To compensate for this, all compressors have an amplifier at the very end where you can crank it back up a bit (*gain*). Sounds contradictory, but what's happening is that once the compressor does its thing, you've got a more controlled

signal with less dynamic level swings. Now you can take this stream-
lined signal and crank it up as needed. You set it in the mix and it won't
jump in and out as much.

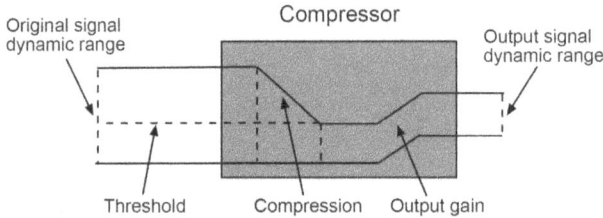

Compressor design and types

A variety of approaches have been developed for compressors, each with
its own particular sound and results. Here are a few concepts you'll
run into.

Optical

The optical compressor is so-called because it employs an electro-lumi-
nescent panel (light bulb) that glows brighter as incoming signal
increases. A photo resistor detects this change and increases impedance,
which then reduces signal gain. Decreasing input level will dim the
panel, decreasing the resistor's impedance, thereby allowing the circuit
to return to unity gain. There's something magical about this particular
approach because it just sounds amazingly musical. The first unit of its
kind, and undoubtably the most famous, was the Teletronix LA2A,
utilizing an optical cell (T4) whose lineage derived from the Cold War
missile programs. You can find them today as hardware reproductions
from Universal Audio or as modeled plugins; they're particularly
fantastic for vocals.

VCA

VCA stands for voltage controlled amplifier, used in many different audio applications. In a compressor, the parameter settings (threshold, ratio, attack and release) are used by a control signal to tell the amplifier how to attenuate the audio. VCA-based units are excellent for fast-acting control; they react quickly to transients and peaks and are usually hard knee (with exceptions). They are not good for general smoothing or shaping of level and volume, but are quite nice for adding a punchy, aggressive sound. Most of your standard compressors fit into this category.

Variable mu

These are tube compressors that feature an incredibly smooth, transparent, and warm sound. Slow to react, they shape a signal over time. An increase in input level will actually increase the ratio along the way. As opposed to other units that feature a tube for color, this design uses the tube itself to change gain. By the way, "mu" means "gain", in case you were wondering. These are popular for mastering because they can provide a final, overall shape to a mix.

FET

Field-effect transistor-based compressors are the opposite of optical and variable mu in that they aren't even close to being transparent and smooth. These units aggressively add punch and color to a track, featuring a super-fast attack time. The Urei 1176 (Universal Audio) is probably the most famous and loved FET compressor ever made and sounds fantastic on snares, guitars, and so on.

Bus

In the 1980s SSL introduced the G-Bus compressor, found on the mix bus of their large format recording consoles. Engineers mixed hundreds of albums on these boards, finding the compressor to be wonderfully transparent with the ability to "glue" a mix together. The idea of a bus compressor is to be subtle and bring a cohesiveness to the entire mix;

you're only looking for a dB or two of GR. They are typically VCA-based because this design provides a lot of control over the signal and works well on summing buses. Another popular model is the API 2500, which can be set for a more aggressive tone with lots of punch. Although the good hardware models are pricey, plugins are available based on these classics. If you're running a DAW, use the main stereo mix track to insert this compressor (master fader in Pro Tools).

Multiband

A standard compressor operates on a signal's entire frequency range. This can cause issues if, for example, there is a lot of low frequency information in the sound. This will cause the compressor to react more aggressively, thereby hitting the mids and highs too much. Multiband compressors divide the frequency spectrum into bands, or regions, such as low, low-mid, high-mid, and high. This concept should be familiar from the EQ chapter. Now the processor is free to examine each band independently, only applying compression as needed in a particular region. Such units have been utilized in the mastering process for decades and are also very popular for live reinforcement. They are equally helpful when mixing in the studio.

Dynamic EQ/compressor hybrid

This recent concept combines the real-time reaction of a compressor with the frequency-selectable precision of an EQ. Dial in the exact frequency regions and bandwidth that need attention, but let the compressor determine when, and how much, should be applied. This is very beneficial in the studio, but can be particularly effective for live reinforcement.

Hardware vs software processing

Many software plugins are modeled on the original hardware processors, such as the LA-2A, 1176LN, Fairchild, and so on. Which is better?

Depends. I might automatically say I'd prefer the gear in the rack, because software just doesn't sound the same. And it doesn't, but there are advantages to the plugs as well, so as always, use your own judgment and pick what works in that situation. Just don't be like a student of mine long ago who freaked out when the plugins stopped working one day. He was desperate for one in particular, so I simply pointed to the processor rack. "Why don't you use the real thing?"

A significant advantage for plugins is that they're much less expensive. A high quality hardware compressor will generally set you back at least several hundred dollars, but a modeled version might be found on sale for $89. Yet another bonus is that you can employ that single plugin multiple times as if you had half a dozen of them in the rack. Can't do that with analog.

How do you plug it in?

You need one compressor channel for each single channel of audio you want to work on. If you want to compress the bass guitar, acoustic guitar, and the lead vocal, you need a compressor for each part. You can, though, put a compressor on a group of parts if they're grouped on the console. So if the vocals are all routed to an aux send or sub-group, then on to the main mix, you can put a single compressor on that sub-group.

Use the insert points on the console. Connect the insert-send on the channel to the compressor input, then route the compressor output back into the insert-return on the same channel. Inserting processors actually re-routes the signal out of the channel to the processor, then back again to continue through the channel. Think of it as literally placing the compressor into the channel strip, say between the EQ and aux sends. The signal is now permanently changed by the processor—the original sound no longer exists.

It's the same concept for DAWs, where the processor is inserted directly into a track and can only affect that part.

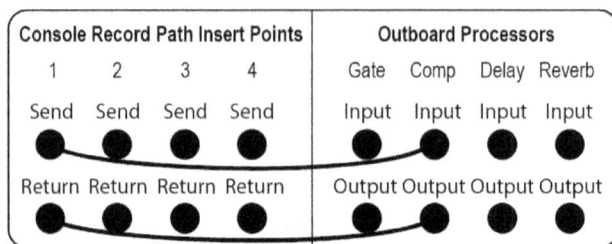

Console Record Path Insert Points				Outboard Processors			
1	2	3	4	Gate	Comp	Delay	Reverb
Send	Send	Send	Send	Input	Input	Input	Input
●	●	●	●	●	●	●	●
Return	Return	Return	Return	Output	Output	Output	Output
●	●	●	●	●	●	●	●

Let's say you want to compress a group of vocal mics all bussed to the same track to be recorded. Normally the multitrack bus out of the console is automatically cabled to the inputs of your multitrack recorder. Simply break this by patching a cable from the console MT bus out to the compressor, then from the output of the compressor into the MT record input. This is the same concept as inserting a compressor into a channel, only in this case you're inserting the processor between the console's output and the recorder's input. Once you understand the signal flow involved, you'll be able to figure out all kinds of ways for applying compression and other processors into your workflow.

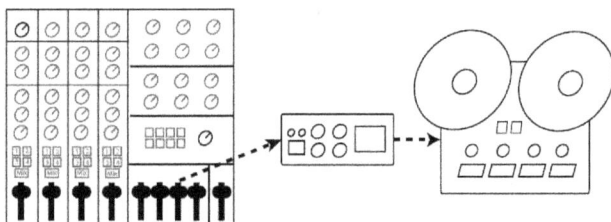

Listen to the following examples to hear what the compressor is doing to the track.

Audio example 102: Compression on saxophone (none, heavy, then medium)

Audio example 103: Compression on acoustic guitar (light, then heavy)

Audio example 104: Drumset with compressor on kick and snare (off, then on)

Audio example 105: Snare (out in)

Audio example 106: Vocal (out in)

Audio example 107: Guitar (out in)

Probably the most noticeable result is the loss of the initial attack. The initial snare hit or guitar strum is the loudest part of the sound; if you have a fast attack setting on the compressor, it immediately holds this back. Listen how it takes a quick dip before leveling off a bit. Of course, different settings will change this somewhat, so put your headphones on and experiment by just turning knobs. Too much compression will diminish the liveliness and dynamics of your mix, but there are certainly times where extreme settings sound really interesting. And don't forget to try different types of compressors—each provides its own flavor.

Notice that the meters will show either input, output, or gain reduction (GR). If the meter sits idle at oVU, it's in GR mode (no reduction taking place); the needle will swing to the left (negative) once compression kicks in.

To get started, try these settings and then experiment a bit for different instruments.

- Ratio: 3:1
- Attack: Keep it fairly fast, but not quite all the way
- Release: Medium
- Threshold: Slowly turn it up or down until you get just a few dB flashing
- Output gain: Start at o (no change) and adjust as needed to provide a good level in your mix. You may not need this if you're only compressing a few dB.

If the compressor meters light up like a Christmas tree, it'll sound squashed. Unless you really want this, raise the threshold. It takes quite a bit of practice, listening, and experimentation to get a feel for setting these parameters, so you might as well start now.

If you are compressing a stereo track, such as two piano mics or a stereo brass section, use a 2-channel compressor and make sure the link switch is on. This locks both channels together so the response is identical, typically using the left channel controls to set the entire unit. If you don't link both channels, each side will react somewhat differently and cause panning shifts, etc.

Lastly, ever notice the side-chain or key input? On plugins, it literally looks like a key icon. Insert the processor on an audio track, feed an external signal into the side-chain (use a bus via send from the source track), and set the unit for key input. It's now looking for something to trigger the compression cycle. The processor is still affecting the sound of that track, but how the processor behaves comes from the other signal. This opens up all kinds of creative possibilities—see the mixdown chapter for several ideas you can try.

You can also do this on an analog console. Use an aux send from the track you want to use as a trigger. Connect this to the key input of the compressor you've inserted on the other track, then set the compressor to key input. Here's the signal flow:

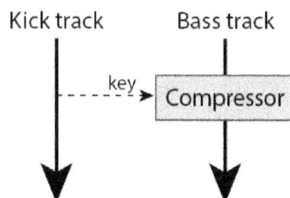

Limiter

Limiters are the same as compressors, only with a high compression ratio (10:1 or higher). The idea is that you want to aggressively control the signal once it goes beyond a certain point, so the threshold is usually high also. While they can be used just like a compressor on a music track, they are also often inserted into sound systems to protect against sudden pops or transients that might damage something (like when someone plugs a microphone in without muting the channel).

Peak limiters are a type of limiter used during the mastering stage. The typical idea is to get as much volume as possible out of the mix so it sounds "just as loud" as other records. It essentially squashes both ends of the signal, bringing high levels down and low levels up. If applied too liberally you end up with a mix that's lost its musicality, ebb and flow, and tires the ears after listening for awhile. Unfortunately, most recordings are processed like this these days; open a commercially recorded track in an audio editor and observe the waveform. It's pretty much maxed out top and bottom with zero headroom. Notice how the following graphic reveals less variation in the waveform along with flattened peaks in the bottom example. Yes, it's louder, which is good, but too much of this isn't so great.

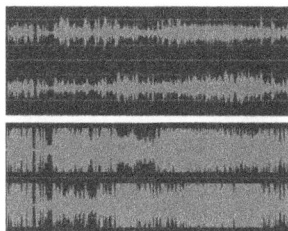

How do you plug it in?

Limiters are channel-specific tools that are inserted directly into a channel or bus on the mixer. Just like with a compressor, connect the insert-send on the channel to the limiter input, then route the limiter output back into the insert-return on the same channel. If used as a mixing/mastering tool, insert the processor on the mix bus or in your stereo audio editor.

Noise Gate

Ever notice background sounds such as a hissing guitar amp or snare drum leakage on the kick track? Sometimes you can reduce this by inserting a *noise gate*, which is designed to attenuate a signal once it falls below threshold. Since these background noises are usually quieter than the main signal you're recording, you can set the gate to drop the signal when the main part is not playing. So for a guitar track, set the gate threshold low enough that it stays open when the guitar is playing, but will shut down (attenuate) while not playing. This will get rid of the amp noise until the music starts. Note that you can't gate the noise when the musician is playing—you have to leave the channel open so you can hear the guitar. However, the idea is that the guitar sound will mask most or all of the background noise, so it's not as big of a deal.

Another term for noise gate is *downward expander*, the idea being that the processor is increasing dynamic range by lowering low level signals. This is opposite of what a compressor does, and can be used not only for reducing unwanted sounds, but also to introduce more dynamic range to the music.

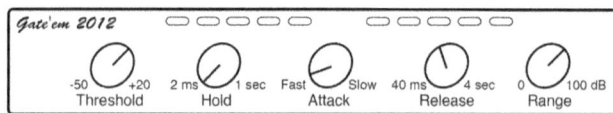

Functions and controls

Just like with the compressor, *ratio* represents input vs output levels—

how much the gate is attenuating. A ratio of 3:1 means that the gate will allow only 1dB out with every 3dB that comes in. You have to keep in mind that a noise gate doesn't actually turn the channel off, but merely attenuates the signal to a point you set. *Threshold* is the point below which the gate will attenuate the channel. As the program material falls below this level the device reduces gain to a user-set *range* or *floor*. This may actually not be all the way down; sometimes it's sufficient to merely drop the signal a few dB rather than try to cut it out altogether. *Attack* and *release* times operate similar to a compressor. Attack determines how quickly the gate will open up once a signal comes in above threshold, while release sets how long it takes to "shut down". You need to play with these settings to find a combination that works for each specific situation; otherwise the gate might chop off sounds too abruptly or kick in too slowly. Some gates also provide a *hold* timer—after the gate has opened, it will hold it open until after this time has elapsed, even if the program material has already dropped below threshold. It's probably a good idea to leave this off while you're learning how to use the other parameters.

How do you plug it in?

Gates are channel-specific tools that are inserted directly into a channel on the mixer, so plug it in just like you did with the compressor. Patch the channel insert-send to the gate input, then from the gate output into the channel insert-return.

Audio example 108: Guitar amplifier hiss with gate off, then on

Audio example 109: Snare with kick leakage—gate off, then on

Noise gates also have side-chain inputs (key in). It functions the same as for compressors, where the external signal triggers the gate cycle. One example is for frequency-selective gating, where you can have the gate respond every time the high hat comes through while ignoring the kick. See the mixdown chapter for ideas.

De-esser

With vocals or speech, the consonant "s" sometimes gets a bit out of control. We call this *sibilance*, and it's prominent around 6kHz or so. You can use a compressor to reduce the effect, but you probably don't want to constantly compress the entire sound. A de-esser will target compression whenever it detects excessive energy in that frequency range.

If you don't have a de-esser, you can set one up with a compressor and extra EQ. Insert a compressor on your vocal channel and set it to external/key mode. Now take a copy of your vocal and run it into a separate EQ (not on the same channel). Boost the 6kHz range on this EQ considerably—crank it way up. Patch the output of this EQ into the key input or side-chain of the compressor; its only function is to trigger the compression and won't be heard in the mix. Now whenever the "s" frequencies occur, they are exaggerated in the EQ circuit, which then tells the compressor to do its thing on the main vocal channel.

De-esser plugins allow you to dial in the exact frequency range that's causing problems and determine how much reduction works for the situation. This is not just for vocals—it works well on guitar string squeaks and similar noises.

How do you plug it in?

Same as compressors, using insert points on the specific channel.

Getting started with compressors

All too often novice engineers avoid compressors because they seem complicated. Take the time, run signals through them, and experiment with the controls to see and hear what they do. Get a feel for various settings with different instruments and voice tracks. Try different compressors to hear how each offers a unique sound and result. The following list of ideas can help guide you as you practice, and remember to review the creative ideas in the mixdown chapter.

Overall concepts

- Compression smooths performance fluctuations, stabilizing a track in the mix so it doesn't jump in and out so much.
- Compressors add sonic character and variety to your sounds. You don't always have to see much, if any, GR.
- Experiment with different settings from gentle to extreme.
- Try different instruments and styles of music.
- Try different compressors.
- Use your ears—there is no "right" setting. Experiment!

Instrument-specific tips

Compression on drums can help tighten and fatten the sound, provide enhanced attack, and make hits more uniform in the mix. You can set either gentle compression or hard-hitting limiting for effect (such as for overheads).

Kick

- Try just a few dB of gain reduction.
- Play with the attack time—too fast and you lose transients, too long and you've missed it completely.
- Use minimal compression to bring up low end without EQ.
- Compare optical and FET compressors to see which gives you the flavor you want: optical is smooth and transparent while FET is more aggressive with punch.

Snare

- Generally same concepts as for a kick track.
- Compression on a snare contributes toward that tight studio sound you hear on records.

Bass guitar

- Helps even out the performance in the mix.
- Extend notes a bit with longer release times.
- Compare optical and FET models and decide whether you like it more smooth and transparent or aggressive and in your face.
- Try compressing a direct and miked signal with different compressors.
- Bus direct and miked tracks together and compress as one signal.
- Use a multiband compressor which provides separate compression circuits for different frequency ranges (low, mid, high). Compress the low-frequencies only; this leaves the mids and highs as-is for better attack and definition.

Piano

- For classical music you probably don't want any.
- For rock, you can use heavier compression to fit the instrument into the mix and give it a more aggressive sound. The piano is often functioning as a rhythm part in this type of music and doesn't usually need to stand out so much.
- If you're stereo miking the piano use a stereo compressor and make sure the *link* switch is on.

Vocals

- Compression can smooth a vocal performance in the mix, especially for non-professionals.

- Experiment with the release time; if it's cycling too quickly you'll get a breathing or pumping effect.
- Apply a de-esser first in the processor chain as needed to reduce sibilance.
- Try various flavor compressors, such as opticals and tubes. Gentle shaping can be quite effective, but sometimes a hard-hitting, aggressive FET with a touch of distortion works really well.
- Try two compressors in a chain, where the first hits the dynamic swings and the second provides overall shaping and color.
- The right combination and settings will help your vocal stand out front in the mix with lots of presence.

Guitars

- Same general notes as we've described so far.
- Rhythm parts can use heavier compression to focus them in a supporting groove.
- Lead and solo acoustic guitar might sound best with less compression, but then again...

Miscellaneous

- Combine kick and bass guitar (or whatever) on a bus and compress together.
- Try the parallel compression technique we discussed in the mixdown chapter. Assign all drum tracks to a stereo bus and compress fairly heavily, combining this result with the original tracks at the mix bus. Very powerful. This technique works well with lots of instruments, not just drums.

SIXTEEN

SIGNAL PROCESSORS—TIME

Time is the final domain we can work with in an audio signal. Sounds have life cycles that take on various attributes over time before they die away. The physical environment influences this significantly, and we can alter or recreate the time envelope of sounds in rooms with time-based processors. A basic foundation of sound behavior in acoustic spaces is necessary for understanding how these work.

Sound generated in a space will spread out and reflect around the various surfaces. Size, shape, and surface treatments determine frequency response and the nature of these reflections, meaning they differentiate a bathroom from a concert hall.

A propagated sound passes through three stages:

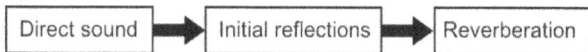

Direct sound → Initial reflections → Reverberation

Direct sound is what you hear straight from the source, with no reflections from nearby surfaces affecting it. The closer you are to a sound source, the more direct sound you will hear.

Initial reflections are the first waves to be reflected from nearby surfaces, primarily the front and side walls. The delay time between the original direct sound and these early reflections provides us with a sense of how large the room is. Think about it—in a larger room the walls are farther away, so it takes longer for a reflection to bounce off the wall and arrive at your seat. Therefore a longer initial delay sounds like a bigger room.

The reflections in a room rapidly multiply, particularly when the source continues to produce sound. As the reflections become more numerous and dense the result is *reverberation*. This evenly distributed, diffuse sound field contains no discernible reflections and will eventually die away at a rate dependent upon the physical attributes of the room (size, wall surface textures, and so on).

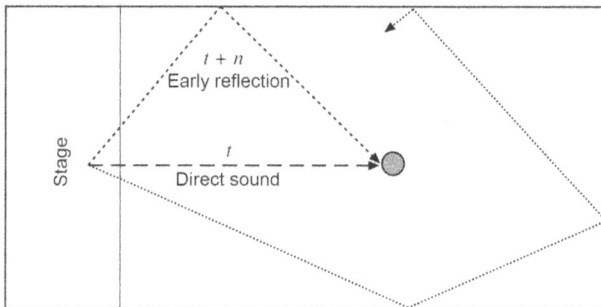

Time-based processors operate on these principles, allowing you to craft any environment you wish. Sometimes this is to merely recreate a familiar acoustic space for a performance, but many tools are used in creative ways to enhance your mix. Let's take a look at the more common processors, find out how they work, and hear what they sound like.

Reverberation

We always hear reverberation of some type and amount wherever we are. You've heard the difference between talking and singing in your living room, bathroom, and the school auditorium. Sometimes in studio recording we try to reduce reverberation in favor of a closer, drier signal.

Then we can artificially add whatever we want for the final mix. Reverberation processors take incoming signals and generate numerous reflections, simulating any particular environment you want. Over the years these are the main types of units we've seen in the studio.

Spring

Ever hear someone drop a guitar amplifier and it went "sproing"? That's an example of a spring reverb, where an actual spring is connected to transducers. Feed a signal into the unit and it travels back and forth through the spring, generating multiple reflections on the output. These aren't used much anymore because they just don't work very well. I had to use a spring reverb many years ago, and every time the snare hit it overloaded the spring. Not very musical.

Plate

A plate reverb unit is a large metal plate suspended under high tension. You set it up in a separate room and feed a signal into the attached transducer; the vibration of the plate generates reflections that are picked up by other transducers attached to the plate. These reflections are then fed back into the console. These also are not common anymore, especially since you can get much of the same effect from a digital processor.

Chamber

Early studios had a separate room dedicated to generating reverberation. The engineer would feed a track to speakers mounted in a highly reverberant room; the sound would bounce around and be picked up by microphones that were then fed back to the console. This isn't done much these days because using a room exclusively for reverb is expensive and not really necessary due to digital processors that do such a great job. But a lot of great records were cut in studios using this approach.

Digital

Although you still find plates and chambers here and there, for the most part we've gone digital. These processors take incoming signals, convert them to digital, then crunch algorithms to achieve a desired effect, such as reverb, delay, or other special sounds. You can control lots of parameters to get it sounding just like you want and then store this as a preset for later use. They're not just for reverb, as you can generate all kinds of acoustic phenomenon and special effects.

DSP (digital signal processing) has come a long way since it was introduced. The early digital units sounded, well, not very good, but over time the algorithms used to simulate room reflections were vastly improved.

Convolution reverb

This is a process where actual samples of a room are recorded, providing an acoustic profile of that particular space. An impulse response is generated in the room, which is a very brief, sharp transient burst (like a starter gun or bursting balloon). The resulting sound in the room is captured and stored as an impulse response model. Reverb is generated by combining an audio track with the samples from the model in a process referred to as convolving, where each sample of the audio track is multiplied with data from the impulse model.

Types of reverb patches (presets)

Halls

Hall patches simulate the environment of an auditorium or other large space. Longer first reflections, attenuated high frequencies, and denser, longer reverberation time are what make these sound distinctive.

Rooms / chambers

Smaller spaces are characterized by short first reflections and less reverberant sound. Think of a club or your living room.

Plates

A plate preset mimics the physical plate units described earlier. These
typically sound brighter and crisper than hall or room settings.

Effects

Most processors not only generate reverb, but also produce lots of
various effects. We'll describe a few of these a bit later.

Digital Reverb Unit

User-set parameters

Reverb time (RT60)

This represents how long it takes reverberation to die away to a point
60dB below its original level. A large hall will have a longer RT60 than
your living room. When mixing you want the RT60 to match the tempo
and style of the song, so a faster song should have a shorter RT60 time;
otherwise the reverb builds too much and muddies the mix. For slower
ballads, a longer RT60 might be appropriate to allow the music to linger
longer.

RT60 is frequency dependent, meaning high frequencies will attenuate
much quicker than lows. It's also directly related to the room itself,
including size and surface materials. Wallace Sabine, considered the
father of architectural acoustics, developed an equation for determining

the relationship between RT60, room size, and the overall absorption coefficient of the space. This enables studio and facility designers to predict how sound will behave in a particular room, even before it's built.

Pre-delay

This is when the first early reflection will occur, based on where the nearest surface is located. Set this lower for smaller rooms, longer for larger rooms.

Initial reflections (early reflections)

After the pre-delay time, the first reflections begin arriving from various wall surfaces and contribute to our sense of how large the room is. Longer instances of early reflections will push the track farther back in the mix. The range provided on effects processors is typically 0-100ms, but 0 is quite weird since there's no such physical space with absolutely no distance between the source and the nearest surface. Try between 10 and 30ms or so to give a vocal some space without losing it in reverb.

Hi/Lo EQ

Most reverb units provide a simple EQ control so you can brighten or darken the reverberant sound.

Wet/dry mix

The output of the processor can either be all effect (wet), all original (dry), or a combination. Since reverbs and effects are most commonly accessed via aux sends, we end up with the original, dry tracks and the reverb return feeding the mix bus separately. In this case the output for the reverb unit needs to be 100% wet. If a verb or effect is inserted directly on an audio track, then a balance needs to be set.

Other settings

Some simple reverb processors only provide a few settings, but others allow you to completely shape how the sound changes over time in a particular type of space. Parameters such as size, shape, and diffusion fine-tune what the reverb sounds like, how long it lasts, exactly how it dies away, and so on. Play around with these various settings and see what you come up with.

How do you plug it in?

Reverb processors are usually not channel-specific devices. They can be connected to the mixer for use on any number of channels through aux sends and returns.

For example, in an analog studio plug the aux 1 send to the reverb unit input, then patch the reverb unit output to any available aux return on the console. On each channel you wish to add reverb, simply turn up the aux 1 pot. Remember signal flow? All these signals on the aux 1 bus combine at the master aux send 1, which you've routed to the reverb unit. The effected sound comes from the unit via the aux return, which sends it directly to the mix bus, not the original tracks. In other words, the two sounds (dry and wet) converge at the mix bus.

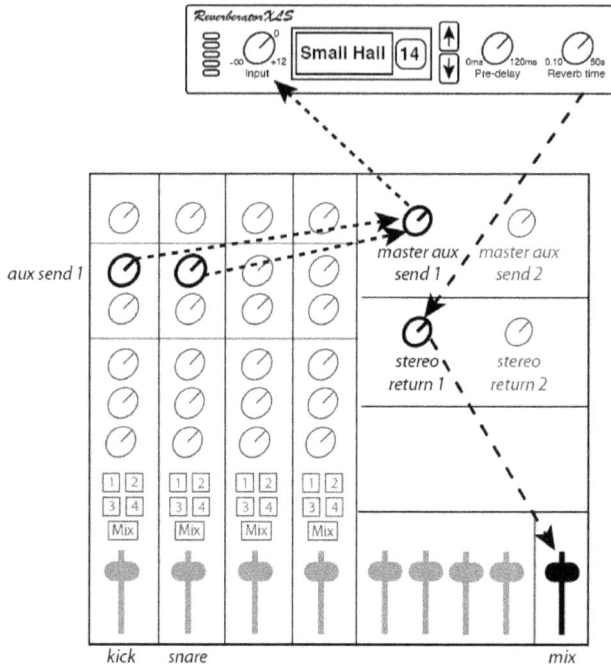

Another option for bringing the reverb back into the console is to use an empty channel. Connect the reverb output to the line inputs of the channel; now you have the capability of EQing the reverb sound, adjust its level with the fader, and even send a copy to another effect chain via a different aux send. Just don't turn up the same aux send that's feeding this device or you'll get an electronic feedback loop.

The typical mistake of many DAW users is to insert the same reverb plugin on each of the tracks. This is not only unnecessary, but it consumes large amounts of DSP power in your computer. Follow the same aux send and return signal flow concept as for an analog console. Use a bus (send) on all the tracks you want to add the same reverb. Select this bus as the input source for an aux track and insert a single instance of the plugin on this track.

Application

Reverb is used for most any recording project. Try different patches to see which is more appropriate to the situation. You will often use more than one processor and patch setting in each song, such as putting a plate on the vocal and a large hall on the drums. Also carefully listen for reverb time; make sure the decay of the reverb matches the tempo and style of the song. Don't just settle for the factory patches—they're meant to be adjusted as needed.

Audio example 110: Small room reverb setting

Audio example 111: Small hall reverb setting

Audio example 112: Plate reverb setting

Audio example 113: Reverb with long RT (2.6s)

Audio example 114: Reverb with short RT (1.0s)

Audio example 115: Reverb with long pre-delay

Audio example 116: Reverb with gradually increasing pre-delay

---

Delays

A delay processor simply takes an incoming signal, holds it for a certain amount of time, then outputs it back to the mix. In the really old days we used tape machines. The physical distance between the recording and playback heads provides a built-in delay, so recording a signal to another machine in the room and monitoring off the playback head returned a signal slightly late compared to the original. This was the origin of the "slapback" echo effect; John Lennon's vocals are a great example of this on the Beatles' recordings. Changing machine speeds also varied the delay time, and if you continuously varied the speed other really interesting things happened. This got easier with tape-based delay devices such as the Echoplex, where the box actually had a continuously-running loop of tape inside. Feed it a signal and the output timing could be adjusted by a slider moving the relative head position.

Digital units input a signal, convert it to digital data if it's analog, store the numbers in a buffer for the desired delay time, then output the result. Sound quality for modern units, including plugins, is far better than the old tape-loop systems, but since engineers are still rather fond of tape-based echo, you can find slap delays on tape emulation plugins complete with that analog tape sound. Guitar players can even get in on the action with a modern Echoplex delay pedal.

Digital Delay Unit

Settings

Delay time

Determines how long the processor will hold the original signal before sending it out. Internally it's making a copy of the original, and the user decides how much of each is heard at the output.

Feedback

Tape slap generates a single instance of an echo. Delay processors provide the capability to loop that echo back to the input of the unit, where it gets delayed in addition to the original signals. This cycle, called regeneration, continues triggering multiple delays.

Modulation

The concept of modulation is to superimpose a signal onto another in order to control some aspect of it. In this case we employ a low frequency oscillator (LFO) to vary the overall wave shape of the audio signal being delayed. This varies the sound of the track in a variety of ways that are best heard than described, but include flanging, chorus, and frequency modulation. Check out the audio examples below for this to make more sense.

Tap

Delay times can be entered manually if you know what you're looking for, but delay processors also provide a button or pad for the user to tap a tempo with your finger or mouse. Play your track and tap to the beat; you'll see the delay time adjust accordingly.

How do you plug it in?

Delay units can be run as channel-specific devices or in an aux send/return configuration. Plug it into the mixer or DAW track using insert points to delay a single track, with the balance between original and delayed signal set on the processor itself. The advantage of using an aux send, though, is that you now have a separate, delayed copy of the track which can be panned, EQ'd, and compressed differently. Route the return into an empty channel for balancing with the original signal. On a DAW simply duplicate the audio track and insert the delay on the second track, set to 100% delay output.

Application

Delay is useful for a number of things. The traditional slap echo is very effective, especially for vocals. Set it for around 60ms and bring the level down a bit; you may not want it calling attention to itself and competing with the main vocal. Another example is running a copy of a single rhythm guitar track (or organ, etc) through a delay, then panning the delay signal different from the original. The dual guitar track, when delayed, provides more depth to the part and the delay helps it move more in the mix. Vocal delays other than slap are employed a great deal, but often in subtle ways you won't even notice. Multiple instances of different delays on a lead vocal might be panned around the main track, but at a low level where you don't really hear them; mute the delays and you'll miss it, turn them back on and they just add a feel to the mix.

Make sure you set the delay times to match the song. This could be quarters, eighths, sixteenths, triplets, or whatever, but make it work musically. Also note that with very short delay times, say less than 15—20ms, if both signals are fairly equal in level and panned together, you'll get severe phasing that changes the overall sound of the track. Comb filtering throughout the frequency spectrum occurs because different frequencies have different physical wavelengths, and if the time relationship is offset between two identical signals, the acoustical summing of the waveforms results in some frequency components getting louder, others softer. Experiment with delay times and pan these away from each other to get a rich, subtle motion effect that works quite well in a mix.

Audio examples 117 & 118: Adding delay to acoustic guitar (out, in, then in context with mix)

Audio example 119: Adding delay to double an acoustic track

Audio example 120: Adding delay in mix context

Audio examples 121—124: Adding delay to guitar track. Original panned left, then delay in right channel at 30ms, 60ms, 120ms

Audio examples 125 & 126: Processing a vocal in the mix, first dry, then w/ delay & verb

Other effects processing options

Doubling

You can make a track sound fatter by generating another copy of it and delaying it very slightly. You won't hear the separate attack, but the double-attack thickens the sound just a bit. Make sure to play with the delay time, since short delays can wreak havoc on the timbre of your track.

Acoustic doubling

A better way for doubling a track is to have the musician record another take on a separate track; since no two performances will ever be identical this results in a richer, thicker sound. Along with the minor timing differences (delay), this method also provides slight intonation differences which contribute to the effect.

Chorus

Ever thought about what makes a choir sound different from a solo singer? All those voices cannot sing exactly the same—there are slight differences in timing, timbre, and intonation. Aside from using a real choir, you can recreate this effect with an effects processor. The *chorus* setting will take the incoming original signal and make a copy. The copy is delayed and slightly detuned, then combined with the original, resulting in a shimmering, fuller sound. Not quite as good as the real thing, but it helps give a track more interest and life. Chorus is often used for background vocals, keyboard pads, organ, guitars, and so on.

Audio examples 127 & 128: Chorus effect

Flanging

Flanging occurs when a copy of a signal is continuously varied in time relationship with the original. Run a track through a short delay, then constantly vary the delay time while listening to both signals. The combination produces an ethereal sound quality. The term *flanging* came from the practice of having two identical tracks on two different tape machines—you start both at the same time, press slightly on one machine's reel (flange) with your hand to slow it down, then release to let it speed back up. You're causing an extreme phasing relationship between both signals.

Flanging is very effective for guitars, vocals, drums, etc. It's a more pronounced effect than chorus. Try experimenting with either flanging or chorus to help keyboard patches such as organ come alive with more depth and movement. Send a keyboard track to the processor, bring the flanged track back into a spare channel, add verb to both channels (so they both have similar ambiance), and pan apart. Remember to EQ similarly also. Or maybe not. Use your ears and play around with it.

Audio examples 129—131: Flanging effect

Phasing

Producing a sound very similar to flanging, *phasing* uses very narrow bandwidth filters (remember the EQ chapter?) that are swept up and down the frequency spectrum. Two copies of the signal are used, one with the filters sweeping up and down, the other not affected. When combined, the filters create phase shift which cancels against the original signal. This cancellation affects different frequencies as the filter moves up and down, producing a very similar ethereal effect as flanging. Very popular and common for guitars.

Audio example 132: Reverb with phaser

Time compression/expansion

Sometimes you need to make a 62 second audio track fit into a 60 second slot, such as for radio commercials. On an analog tape recorder, changing playback speed also changes the pitch—faster increases pitch, slower lowers it. Extreme variations will get you either chipmunks or a slightly dazed Darth Vader. You might get away with it for short adjustments, but with DSP it's possible to recalculate audio data and adjust time duration without affecting pitch. Trying to do too much with this can still introduce weird digital artifacts, but it's much better with recent algorithms.

SEVENTEEN

ANALOG TAPE RECORDING

Analog recording on tape is based on the age-old principle of basic magnetism. There is a rock called lodestone that attracts iron particles; once we figured out this wasn't some sort of voodoo, we began calling it a *magnet*. It turns out that certain alloys can be magnetized for long-term retention, others more temporary. So we can create artificial magnets by winding a coil of wire around a bar of metal and then sending a current through the wire; this magnetizes the metal, producing a magnetic field. If it's a hard alloy the magnetism "sticks", which is perfect for recording tape. Soft alloys, on the other hand, are ideal for tape recorder heads that pass magnetic flux energy to and from recording tape. We don't want the magnetism to be retained in the heads, otherwise it would negatively affect the signal being recorded or reproduced.

Analog recording works by sending an electrical audio signal to a recorder. This is transduced into magnetic flux energy in the record head, which has gaps for the magnetism to flow across. As the flux bridges this gap, it magnetizes the tape as it passes by. Recording tape features a foundational layer, such as acetate or polyester, coated with ferric oxide particles. These particles are what get magnetized, and when the tape passes by the flux field it orients the particles in a pattern of north and south analogous to the positive/negative audio signal sent to the recorder. The goal is to orient as many of these particles as possible; any remaining random particles represent rapid fluctuations in north-south orientation. This is what causes tape hiss, one of the banes of analog recording.

For playback, also known as *reproduce*, the opposite process occurs. The previously magnetized tape particles flow past the playback head. Flux enters the gap, where it's then transduced into an analogous electrical signal.

Magnetic recording tape

The ferric oxide particles found on recording tape are very small, less than 0.5 micrometers in size. Each particle contains at least one domain, which is the smallest physical magnet found in nature. The goal in recording is to orient these particles in a pattern that is linear and faithful to the original signal—you want whatever you record to come back the same way. The challenge with magnetizing these domains is that they suffer from inertia—they don't want to move unless enough force is applied, and the resulting signal is a non-linear (inaccurate) representation of the original sound. This requires a couple of approaches, the first being stronger recording levels. The higher the recording flux, the more particles are aligned and retain their magnetism. Of course, if you go too far all the particles are oriented. This is known as *saturation*, and anything beyond that sounds distorted. But before you reach nasty distortion, as the tape begins to saturate we get a very nice, fat compression sound due to increasing third harmonic distortion. Engineers use this to great effect by recording drums and guitars at elevated levels. You've got to know your gear, though, to find the happy medium between too low (hiss) and too high (distortion).

The other component for enticing magnetic particles to move accurately in response to the incoming flux field is called *bias*. This is a very high frequency tone (150kHz or so) that is superimposed with the audio signal during recording. High frequencies feature extremely short wavelengths, meaning they fluctuate very rapidly. This fluctuation helps overcome the non-linearity of the tape's response, meaning it gets the particles up and moving into position more effectively. Upon playback, the tone is not heard because it's beyond the range of hearing and gets filtered out by equipment that cannot reproduce such high frequencies. In practice, the engineer sets the bias level on the recorder, and this must be done for each type (brand) of tape formulation used in the studio. The recorder's manual will detail how to set bias for that particular machine, but the basic idea is that as you increase bias level, the signal being recorded becomes more linear (accurate). But if you add too much bias it begins erasing high frequencies. So it's a compromise. Many engineers will intentionally set bias differently than recommended in order to get a certain sound for a recording. That's where knowing your gear comes into play!

Both flux level and bias are crucial for an accurate recording and must be carefully calibrated on the tape recorder. This also changes with different types of tape from different manufacturers. However, these are only two of many parameters that must be checked by the engineer on a regular basis. We'll discuss those in a moment.

Magnetic tape properties

Non-linearity

As we've just seen, it takes effort to get the particles aligned and willing to stay in place for retaining an audio signal. This is referred to as non-linearity, meaning that the force required to magnetize particles does not follow a linearly increasing level. At low levels the particles simply don't respond well and fall back into their random state. With enough additional force, they will align and retain their orientation fairly well. At a certain point the particles are saturated and cannot retain additional signal.

Remanent magnetism

This is how much signal remains on the tape after it has passed the recording head flux field. It is measured in *gauss*. A strong, linear response is required for optimum signal-to-noise ratio and accurate reproduction upon playback.

Retentivity

Each tape formulation has a certain ability to store magnetic energy (think "retention" of flux).

Coercivity

A recorder must erase any existing signal in order to record something else in its place. The amount of flux force required to completely demagnetize the tape is its coercivity. I once had an old MCI 16-track that had problems with this...we'd record something over an old tape, and you'd hear the previous track ghosting in the background. Annoying.

Headroom

When recording on analog tape, the goal is to record as high as possible without overloading (distortion). This usually means targeting somewhere between zero and +3 on the VU meters. Anything between zero (nominal) and maximum is a system's available headroom.

Sensitivity

Each tape formulation will have a relative output level as compared to a standard reference test tape, meaning some tapes will output "hotter" than others. Over the decades, manufacturers developed better formulations with higher flux densities, resulting in higher output tapes with less hiss.

Tape recorders

A professional tape recorder is a complex piece of equipment with numerous motors, brakes, rollers, record/reproduce heads, and electronics. All of this must be calibrated and maintained on a regular basis, which is fairly time consuming and requires very specific expertise. It's also expensive; high quality 24-track recorders cost anywhere from $50,000 to $150,000 back in the 1980's. They're not manufactured anymore, which means that if a studio wants to run tape (and many do), they must locate a used machine that's been thoroughly serviced. Lots of details can go wrong: worn heads, worn transport (motors, brakes), old electrical components, and so on. It's not for the faint of heart, so do your homework.

The primary functions of a tape machine are to 1) run tape by the heads at a constant speed and 2) ensure that what goes in comes out the same way. The first item is accomplished through the physical transport, consisting of a supply reel, take-up reel, guide rollers, tension arms, pinch roller, and capstan. The pinch roller and capstan are the crucial components; when you press play, the pinch roller pushes up against the capstan, thereby pulling the tape past the heads. The capstan is attached to a motor located below the deck surface, and this is what determines the exact speed the tape will run. The other transport items exist solely to feed tape, spool it onto the next reel, and make sure it goes smoothly past the heads. Anything in the chain that affects this will cause speed variations, known as wow (slow change) and flutter (rapid fluctuations). In the analog world, speed derivations will change pitch, so it'll be obvious something is wrong. Knowing what to fix might not be.

The second goal, ensuring that what you record plays back accurately, is primarily an electronic issue. As we discussed earlier, the very nature of analog recording is non-linear, so we have to jump through some hoops to get the thing to work at all. This includes a number of parameters that must be carefully set: record level and EQ, reproduce level and EQ, bias adjustment, and even the exact alignment of the physical heads. Most of these must be tweaked daily before each session, and if the machine is to be operated at different tape speeds, all of this requires calibrating for each.

Bias is used for optimizing linearity of the magnetic domains. Recording level controls how high a flux level will be recorded on tape. With better tape formulations over the years, engineers typically ran their machines hotter to take advantage of the improved signal-to-noise ratios. When calibrating the machine, the desired operating level would be selected, and then both record and reproduce levels would be set accordingly. The user doesn't really see this on the surface when recording during a session; you still want to target a little over 0VU on the meter. But internally that 0VU might represent a very high recording level. Again, it goes back to always knowing your gear inside and out.

Record and reproduce EQ take a bit more explanation. This has nothing to do with desired frequency response of the recording, such as making it sound a bit brighter with some high frequency EQ boost. What happens is that due to the nature of sound, high and low frequencies sound lower in level than the mid-range—the overall sound energy distribution is lower on either end. If that were to be recorded on tape as-is, then the lows and highs would be stored at a lower flux level, closer to the noise floor. We want to record everything as high as possible without overloading, so recorders are designed to artificially boost the highs and lows during recording, but then reciprocally reduce them upon playback. This allows us to get a better signal level on either end of the spectrum. Again, the engineer doesn't see (or hear) any of this during a session—hopefully you just record stuff and hear it back the same way.

Headstacks

There are three heads on a professional recorder: erase, synchronization (record), and reproduce (playback). The erase head is on the left as you look at the machine, and its job is to get rid of any remaining signal from previous recordings. It uses the bias generator to apply a high level signal to the tape, completely saturating it. As the tape moves away from the erase head, the remanent magnetism diminishes to zero, leaving it completely demagnetized. The sync head is in the middle and is designed to play back previously recorded tracks while recording new ones. This was Les Paul's ingenious idea, and modern multitrack recording exists because of him (many thanks, Uncle Les). The third head is optimized for playback only and is used for mixdown.

The reason a different head is required for high quality playback is a relationship between the head gap and frequency response. The playback head must generate an output voltage based on incoming magnetic flux changes. These changes double per every doubling of frequency, and so the output voltage doubles as well, at a rate of 6dB per octave according to fundamental properties of decibels. This continues until the wavelength becomes equal to twice the width of the head gap; at this point maximum voltage output has been achieved. Once the wavelength becomes equal to the gap, scanning loss occurs. Running higher tape speeds extends this upper frequency limit. At a higher tape speed, more physical tape is being used to record the same signal, thus resulting in a longer physical wavelength on tape. So the frequency limit that matches the head gap is higher. Record heads, on the other hand, are best with larger gaps, so when you switch modes on the recorder between sync and repro, you're selecting the head appropriate for the job.

Each headstack must be carefully positioned to match the tape path; this is best left to someone who knows what they're doing. Azimuth is the only adjustment that an everyday engineer might tackle when aligning their machine.

- *Azimuth*: Clockwise/counterclockwise rotation of the head
- *Zenith*: Tilt of the head toward or away from the tape
- *Rack*: Pressure at which the tape pushes against the head
- *Height*: Vertical alignment with the tape path
- *Wrap*: How closely the tape follows the curved contour of the head

Tapes, tracks, and other details

Tape machines come in terms of tape width: 2", 1", 1/2", and 1/4". The standard 24-track recorder found in studios was a 2" machine, which meant it recorded 24 "lanes" of audio tracks along a 2" wide tape. You can't see these tracks by looking at the tape, but you can see the gaps on the record head itself to get an idea. Most 1" machines record 16 tracks, 1/2" is for 8 tracks, while 1/4" tape is used for stereo mixdown. There are exceptions to this, such as using a 1/2" machine for stereo mixing (two tracks, left & right). You can't, however, record 24 tracks on a 2" machine and then try to play those back on a 2" machine set up for 32 tracks. The "lanes" must match.

Machine speeds refer to how fast the tape runs past the heads, starting at 30 inches per second on the high end, then dividing by two down the line: 15 ips, 7 1/2 ips, 3 3/4 ips, and 1 7/8 ips for cassettes. Higher speeds provide a better high frequency response with lower tape hiss and noise. It's also more expensive since you're running through a lot more tape. A standard reel of tape running at 30 ips provides about 16 minutes 30 seconds of run time. At a price of over $300 for 2" reels, this adds up.

Since tape is expensive, you don't want to handle it any more than necessary. Always add a section of *leader tape* to the beginning and end of the reel. This is a white, plastic material that's useful for threading on the machine (loading the tape) as well as inserting silence between songs. Most leader tape comes with timing marks every 7 1/2 inches; keeping the different tape speeds in mind, it's easy to calculate how much leader is needed to insert three seconds between songs. So, let's say the recorder is running at 30 ips, and the producer wants three seconds of silence. I'll let you do the math.

Analog recorders operate in *modes*: input, synchronous record, and reproduce. When the machine is in input mode, you'll see signals bussed from the console on the recorder's meters, but you won't see or hear anything if you play the tape. This mode is useful for getting initial record levels and calibrating the recorder. Once you start overdubbing, though, you need to be in synchronous mode (sync) so you can record and play back what you've already done. This monitors everything from the sync head, meaning it plays back previously recorded tracks while feeding the incoming signal for new tracks to your monitor path. I usually just set the machine into sync mode every time I record or over-dub. For mixing, though, set the recorder to repro mode, which uses the third head for playback. This head is optimized for reproduction and provides a higher quality sound than listening off the sync head.

Most multitrack recorders provide an elaborate remote control called an *autolocator*. This device provides basic transport controls (play, rewind, ff, stop, record), counter indicators, track arming buttons, mode selection, memory locations, jog wheel, and a number keypad for dialing in locations you want to wind the tape. Everything is controlled from here, so most studios shove the recorder into a corner or even into a separate room (to contain the machine noise).

We briefly mentioned setting desired record levels when calibrating a recorder and that 0VU on a meter might represent a much higher than standard recording level on the tape. This is referred to as running *elevated levels* and was made possible with better formulations of tape over the years. Flux levels are measured in nWb/m (nanoWebers/meter). The following chart indicates different operating levels referenced to odB @ 185nWb/m. One of the more common operating levels in use was 370nWb/m, which is 6dB above standard. Engineers simply called this running at +6. Once the machine was calibrated to this internal flux level, you targeted 0VU as usual during recording. Inside, though, it was recording hotter than standard levels, so signals were closer to saturation. Once again, know your gear and its limits. How high can you push the level beyond 0 before distortion? Find out by experimenting.

Desired Level		185 nWb/m	200 nWb/m	250 nWb/m	370 nWb/m
(+9)	520 nWb/m	(-9) VU	(-8) VU	(-6) VU	(-3) VU
(+8)	465 nWb/m	(-8) VU	(-7) VU	(-5) VU	(-2) VU
(+7)	415 nWb/m	(-7) VU	(-6) VU	(-4) VU	(-1) VU
(+6)	370 nWb/m	(-6) VU	(-5) VU	(-3) VU	(0) VU
(+5)	330 nWb/m	(-5) VU	(-4) VU	(-2) VU	(+1) VU
(+4)	295 nWb/m	(-4) VU	(-3) VU	(-1) VU	(+2) VU
(+3)	250 nWb/m	(-3) VU	(-2) VU	(0) VU	(+3) VU
MRL (0)	200 nWb/m	(+1) VU	(0) VU	(+2) VU	(+5) VU
Ampex (0)	185 nWb/m	(0) VU	(+1) VU	(+3) VU	(+6) VU

A day in the life (of an engineer)

Cleaning

Well before the session starts, turn on the recorder so it can warm up awhile. First thing is to clean the heads and tape path. Magnetic tape will shed, meaning the oxide particles rub off as it slides past the heads and roller guides. Keeping this in mind, we generally try to limit how much we run the tape as eventually you'll start to lose high frequencies. These days we'll record a track to tape, then transfer to Pro Tools for further overdubbing and mixing. But back in the day we had no choice. The other issue is that these particles will collect on the record and reproduce heads, clogging the gaps. Audio quality will suffer, so it's crucial to keep the machine clean throughout the day. I'd clean first thing in the morning, then during every break. Use a very pure alcohol or specially designed tape recorder cleaner and apply with lint-free cleaning cloths or cotton swabs. For the heads, rub the swab laterally across the heads in the direction of the gaps, not up and down. This ensures you get everything out of the tape path. Make sure you rub down the pinch roller and other guides really well also.

Calibration

You should check the calibration of the machine on a regular basis, perhaps even daily depending on how well it holds alignment. A full calibration involves a specially manufactured reference test tape, a blank reel of recording tape, a test tone generator, dB signal meter, demagnetizer, magnetometer, and a small screwdriver. Here's an overview to give you an idea of the process:

Once the tape path is clean, check for any remanent magnetism (gauss) using the magnetometer. If it registers flux, turn the recorder off, plug in the demagnetizer, and very slowly move the demagnetizer close in to each head stack, then slowly pull it away. Any sudden move can permanently magnetize the heads, which would be a very bad thing. When finished, turn off the demagnetizer and power the recorder back on.

Reproduce calibration. First thing is to ensure that the machine is reproducing signal accurately. After you've cleaned the tape path, put on the reference tape and rewind to the beginning. Set the recorder to repro mode (playback) and play the various test tones on the tape. It will have 1k for setting playback level, 10k for high frequency adjust, and 100Hz for low frequency adjust. By turning the appropriate pots on the channel cards in the machine with the screwdriver, set each of these to the desired flux level you want the machine to operate. Looking at the chart above, if you want to run at +6 and your test tape is 200nWb/m, set the level to -5VU.

Now put on a blank reel of tape. Keep the recorder in reproduce mode (meaning the meters will show what's recorded, not what's going into the machine). Arm all tracks and press play and record. Using a tone generator, feed a 1k tone to all tracks and adjust the record level pots on all channels to 0VU (not -5 for this step). Do the same for 10k and 100Hz. Bias adjust depends on the recorder, and the manual will provide instructions on how to set it. Usually this means feeding in a 16k signal, turning up the bias pot on the machine until the meter hits maximum, then continuing to turn either clockwise or counterclockwise until the meter falls about 3 dB or so.

There are other adjustments, including setting the physical alignment of the heads. These are best left to the professionals, so if you see screw adjustments on or near the head stacks, leave them alone.

Threading tape

If your project tape is new, put it on the supply reel (left side of the machine) and splice on some leader tape. Thread this through the tape path over to the take-up reel (on the right) and reset your machine counter to zero (timing indicator). If it's a reel you've already been using, it should have been stored *tails out*, meaning it was rewound to the beginning after the previous session. In this case, put the reel on the right side of the machine, thread back to the supply reel, then rewind the entire tape to the beginning. This sounds counterintuitive, but the issue goes back to the nature of recording tape. It stores signal with magnetic flux, which is a field. Anything adjacent will be affected to some degree, and the layers of a tape wound tightly on a reel are in very close proximity. Flux energy will seep into the next layer. You can't avoid this totally, but if the tape was stored heads out, then upon playback the adjacent layer that got affected would pass the repro head *before* the main audio signal. This means you hear a ghost "preview" of the sound a split second before you hear the main sound. By storing the tape tails out, the affected layer comes *after* the main sound passes the reproduce head, so at least it gets somewhat masked by the continuing sound of the recording. You probably need to think about this one awhile for it to make sense; most likely it won't become clear until you actually run tape someday. Meantime don't let it keep you up at night and always remember to store your tapes tails out.

Here's a helpful tip for rewinding a tape you just put on the machine.

Don't hit rewind and just stand there waiting. All recorders have transport counters, meaning you can reset a counter to zero and see how much tape goes past in terms of minutes and seconds. If you're running at 30 ips, you'll get about 16 minutes and 30 seconds or so of playing time. Type in -16:25 in the autolocator, then press *locate*. Now you're free to go set up mics or something while the machine automatically stops at that location (the negative value is because it's rewinding from zero, not advancing). You'll look like a genius and won't have to rethread the tape if you stupidly watch it fly off at the other end.

Recording, overdubbing, and mixing

To get recording levels, place the recorder into *sync mode* and arm the tracks you want to record first. Bus your mic signals from the console and set levels with the mic preamp and channel fader, aiming for around 0VU or a bit higher. Once you've completed a take, turn off the track arm buttons (so you don't accidentally record over what you just got... there is no *undo* in analog), rewind, and press play. Be aware that when the machine is playing a tape, the incoming mic signals are no longer active through the machine, meaning you can't hear the musicians talking to you. Upon stopping playback, however, you want the machine set to *auto input*, meaning that all inputs are active so you can hear any mics routed through the recorder. This gets complicated to explain unless you've actually got a machine to work with, but just know that there are ways to ensure you can hear the musicians talk to you between takes.

Musicians aren't always perfect, and so you'll have to fix a portion of a take. *Punching in* is the process for playing through part of a take and re-recording a segment. The idea is to start playback a bit before the section you want to fix, then punch the record button a split second *before* the precise moment you want to start fixing. The machine will drop into record mode, record whatever the musician is currently playing, then you punch the play button to drop out of record. You've now essentially inserted a replacement section into the track. With a DAW you simply designate the in and out points, and if you don't get it right you click undo, adjust the points, and try again. With analog you've got to get it right. The first time. Or you'll have to get on the talkback to the musician "Sorry, uh, I missed it. Can we take it again a bit earlier in the track please?" I worked with a machine once that actually muted audio when you punched out of record, so you couldn't hear whether the punch worked. Eventually we got pretty good at knowing whether it "felt right", but to be sure you had to take time to play the transition.

For mixing, clean the tape path and put the multitrack recorder into repro mode. All tracks are routed through the console's main signal paths and over to the mix bus. Locate the tape at the beginning of each song and note the counter time; the recorder can be set to automatically return to this point every time you want to start again from the beginning. If you're mixing to an analog two-track, clean it, thread a blank reel of tape, and run the hottest section of the mix to set a good record level. The two-track should already be connected straight from the console's mix bus, so everything you do on the board, including faders, signal processing, and effects, will be mixed and routed to the mixdown recorder. Don't bother trying to time the 2-track so it starts at exactly the beginning of the song; you'll deal with this in editing. Just start the 2-track recorder, then the multitrack.

Editing

Once mixing is done, you've got to go through and remove all count-offs before each song and any extra time and noise after each song. The easiest way is to transfer your mixes back to your editing software, but if you want to go old-school grab a grease pencil, razor blade, splicing block, splicing tape, and leader tape. Put the 2-track recorder into edit mode, which allows you to rock the reels back and forth. This means you grab both reels and turn them back and forth, moving the tape across the heads at slow speed. You'll hear a weird sound coming from the monitors, but eventually you can make out the drum clicks from the count off and the downbeat of the song. Move the tape until it's just barely before the downbeat, then make a small mark on the tape right next to the repro head with the pencil. That's your cut location, so unspool the tape so you can drape it across the splicing block. Push it in so it's snug, then take the razor blade and make a clean cut in the splicing block groove. It takes practice to keep the tape secure while you cut; it's got to be clean or the edit might pop during playback. Measure the needed length of leader tape and cut each end to the same angle in the block. Butt it up against the edge of your song downbeat so it's flush, then apply a short piece of splicing tape. Make sure the edges of the splicing tape don't overlap the edges of the recording tape or it'll stick while the tape is played or rewound. Use your fingernail and rub the entire area of the splicing tape until it turns dark in color; this ensures a solid adhesion. Now, find out whether you passed the test. Thread the tape back on the machine and press play, then rewind and fast forward. If it comes apart, just try taping it again. You'll get the hang of it.

It's actually fun to cut and splice tape, although I'd prefer not to do an entire album ever again. I spent many a long evening editing and sequencing albums, meaning each song had to be edited separately, stored on different reels, then re-assembled in the order they wanted for the album (mixing is done as is convenient for the tracks, not according to the final album order). And then there were the dance show directors who brought me piles of CDs, cassettes, and records for producing performance tapes for their shows. That was great fun...4 bars from this tune, then 8 bars from another, and so on, regardless of what key or time signature they were in. I had piles of tiny slivers of tape lying around the machine, each waiting for its proper place in the final edit, and each in great danger of me sneezing and blowing them to kingdom come all over the room.

Documentation and labeling

Either keep all the take and track sheets inside the 2" boxes, or compile everything in a folder labeled to match the tapes. Every tape must be clearly labeled; you don't want to sort through stacks of boxes looking for that perfect mix the client is screaming for. There was once an album project where an engineer was looking for a blank reel, found one, and proceeded to record over tracks completed the previous week. Here's an example of a tape label with most everything you need:

Client: Project #:
Engineer: Date:

Tape speed: ☐ 30 ips   ☐ 15 ips   ☐ 7 1/2 ips

Tracks: ☐ Mono   ☐ Stereo   ☐ 4 tr

Wound: ☐ Tails out   ☐ Heads out   NR ☐

Track list   Time start

Hear Real Good Studios
101 N Audible Ave
Deafness, ZM 90909
909-555-9090

What's the point of tape these days?

Digital audio technology provides so many benefits for music production, such as super clean, high resolution audio with unlimited editing and processing tools. But there's something about analog that we can't help but love. It's a warm, full sound, and when the machines are dialed in right it's hard to beat. Most studios and audio engineers don't need it, can't afford to deal with it, and aren't interested. But many do, and so there's a healthy trade for old analog machines. You can even buy new blank tape, currently manufactured in York, PA. Usually the idea is to record basics and overdubs to tape, then transfer immediately to Pro Tools for further editing and mixing. It's the best of both worlds.

EIGHTEEN

NOISE REDUCTION

One of the major drawbacks of analog recording is tape hiss, an audible high frequency noise. I explained this in more detail in the analog tape recording chapter, but what's happening is that the magnetic particles have to be aligned in order to store a signal. It's nearly impossible to get them all, so any randomly oriented particles cause rapid fluctuation in the resulting signal upon playback. Faster tape speeds are better because doubling the speed also doubles the frequency range of the hiss, raising much of it out of the range of human hearing. The worst case is the plain old cassette, which slogs along at 1 7/8" per second.

Analog tape has a finite amount of "room" in which to store signals. If you record too low the noise floor becomes equal with the signal. Record too high and it saturates the tape, causing distortion. The dynamic range in between is very limited in this medium, much less than normal music performance; digital recording has a much larger range, so it's not really an issue even with the most expressive sources such as orchestras.

There are a few tricks to reducing tape hiss such as recording at higher tape speeds, recording as hot as possible without distorting, and even boosting high frequency EQ during tracking with a reciprocal reduction during mixdown (which reduces the hiss along with the audio). None of these completely solves the problem, however, so a more technical approach is needed.

The idea is to compress the audio signal being recorded, then expand it back to normal during playback. The basic process is *compansion* (compression and expansion), but we refer to it as *noise reduction*. Compansion works by raising low-level signals during record. When these are attenuated back to normal during playback, the tape hiss added during recording is lowered as well. Since this noise is a fairly low-level signal to begin with, the resulting lower level effectively reduces the amount of hiss we hear. Different systems approach this in a variety of ways. The old dbx compansion process simply companded the entire frequency range. More sophisticated systems from Dolby broke the spectrum into ranges that were dealt with as needed.

Types of noise reduction

Dolby A (*professional*)

Ray Dolby's Type A system was an early professional NR unit for studios. It divides the frequency spectrum into four fixed frequency regions. When a signal falls below threshold in any region, gain is increased using a sidechain procedure. This is reciprocated during playback and provides up to 10dB of noise reduction below 5kHz. This is essentially the compansion process, but operates individually in separate frequency bands.

The sidechain works by passing the original signal in each band untouched. A copy of each band runs through a limiter; when the input signal for a band is low, there is little limiting performed. This uncompressed level is then added to the original, resulting in a level gain. If a band's input signal is high, the sidechain limits quite a bit. This lower signal doesn't add as much to the original when combined, so the resulting output doesn't change much. What you end up with is that the low-level inputs have a higher gain output from the process during recording, which is essentially the same idea as basic compansion—low level signals are raised during recording, then balanced back to normal during playback. Dolby A is simply a bit more sophisticated by breaking the signal into different frequency bands.

Dolby SR (professional)

In the mid-1980's Dolby SR arrived and seemed like magic to those of us trying it for the first time. Also working on the sidechain principle, SR features a complex division of frequency regions, performing a real-time spectral analysis to determine what areas need gain changes. Instead of four fixed frequency regions as in Dolby A, there are five groups of fixed and sliding band filters. These can be assigned on the fly by the analysis circuits to treat any problem areas. The result was an impressive 24dB improvement.

Dolby B (consumer)

Nearly every cassette recorder manufactured from the 1970s onward featured Dolby B for record and playback. Commercially produced cassettes bought at the record store were Dolby encoded. This system featured a single fixed-gain, variable bandwidth circuit that boosted highs during recording and reduced during playback. The result was about 10dB of extra signal-to-noise. If a cassette player did not have a Dolby circuit, it would still sound good, just a bit brighter. Often people would switch Dolby off during playback to get that brighter sound.

Dolby C (consumer/semi-professional)

The successor to Dolby B was essentially two Dolby B circuits in tandem—first a high level, then a low level compansion circuit resulting in 20dB noise improvement. It did not sound good if the player had no Type C circuit for decoding and never really caught on.

Dolby S (consumer)

This was based on the professional SR system and intended for the consumer market. Since Dolby B was the absolute standard, Type S never caught on to replace it. Some high end cassette recorders included Dolby S.

dbx (professional and consumer)

dbx introduced their Type I & II NR systems in the 1970s. It featured a straightforward 2:1, single-band compansion process with high frequency pre-emphasis. The drawback to companding the entire frequency range at once is that it continuously changes gain across the entire sound; it's not restricted to the high frequencies. The result is an audible breathing effect, especially if the music performance has a wide dynamic range. It never caught on in the consumer market; Dolby B was already in widespread use, and it did not work on non-dbx equipment. For the so-called semi-professional market that flourished in the 1980's, however, it found an eager user base. The popular Tascam Portastudios and other small-format recorders employed dbx quite successfully.

Single-ended NR (professional)

This is not a compansion system of code and encode, but rather a frequency selective downward expander (like a noise gate). It's only inserted during playback and attenuates high frequencies as set by the user. Of course this grabs your audio along with the hiss, so it's a careful balance between noise and impacting high frequency response.

NINETEEN

DIGITAL RECORDING

Sound is an analog event, but there are numerous advantages for encoding this as digital data. Analog recording is inherently non-linear and chock full of issues as we discussed in the analog recording chapter. Digital information, on the other hand, can be stored, transmitted, and altered with little or no degradation of sound quality. At its core, digital is simply a string of zeros and ones, known as binary. This can be represented as on or off, so the system only needs to distinguish between two values. Incredibly, such complex information as audio signals, metadata, error correction, and so on can be generated by combining strings of these two digits into meaningful digital words. The theory has been around for generations, though it took a while to develop the process and hardware so it sounds as good as, and in some ways better than, analog.

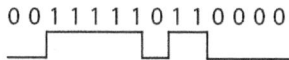

0 0 1 1 1 1 1 0 1 1 0 0 0 0

Sony released the first commercial pulse code modulation recorder in 1977. I'm not quite that old, but while in college in the late 80s we had the improved version of that model, the PCM-F1. And what a system it was! Audio fed into the processor unit was converted into digital PCM data; this was then stored onto a standard Sony Beta video tape recorder. Of course you couldn't grab the tape and run out to the car to check your mix, as it was only playable on the PCM. But we could edit, sort of. The video display (no color, of course) showed an image that looked like a sonar "waterfall" display, where lines or waves of static move across the screen. Some areas of this noise would be tightly bunched together, marking a transient. You couldn't hear a thing, but once you got the hang of it you could tell snare hits and downbeats. We'd set an edit in-point, find the out-point, then pressed record for the system to make the corrected recording to a second VCR. No undo, no audio, no confidence. But it was digital editing.

Over the years the industry developed a variety of digital formats and storage media, including the washing-machine sized DASH tape recorders, tiny DAT videotapes, and of course the ubiquitous compact disc. Some sounded better than others, and generally speaking as time went on digital systems got much better and more appealing to musicians and engineers. In the 1990s the concept of perceptual coding came in the form of Mini-Disc and Digital Compact Cassette, two competing formats designed with consumers in mind. The idea was to reduce data, making it possible to store music on portable media with limited capacity. Perceptual coding attempted to identify aspects of a musical signal that were masked within the mix and therefore inaudible to a listener. It worked, mostly, but the idea was so revolting to the music industry, and confusing to the public, that those formats never caught on. They were also rendered essentially irrelevant with the development of the recordable compact disc (CDR), which let consumers make their own recordings at a quality far superior to the analog cassette.

Today, of course, everybody listens to perceptually coded music, primarily as mp3 or AAC (advanced audio coding). These are known as *lossy* formats, meaning the data compression techniques throw away actual audio information in the pursuit of small file sizes. We're not talking about dynamic range compression, but rather how a computer manages data files. There are also *lossless* formats, such as FLAC (free lossless audio codec) and ALAC (Apple lossless audio codec), that reduce file size as much as possible while preserving original audio information. Naturally a lossy format will give you a much smaller file size, and since network bandwidth and portable device storage was rather limited for years this became the practical, widely adopted solution.

Raw digital audio is usually recorded as WAV (WAVeform audio format) or AIFF (audio interchange file format) files. These utilize no compression whatsoever, capturing the original analog sound as–is. File sizes are large, depending on the resolution used, but all the content is there. Engineers always record as high a resolution as feasible, then work downward depending on the ultimate application. If you've ever tried enlarging a low resolution jpg photo, the pixelation jaggies become obvious. Same with audio—always start as high as possible.

So exactly how is an audio signal encoded? We'll simplify it somewhat here, understanding that it's a fairly detailed, complicated process underneath the hood. First, remember that an analog sound is a continually changing event. Compare an analog clock with a digital one, where the second-hand smoothly rotates around the clock face. A digital clock will increment in steps. For a digital recording we have to convert an ever-changing signal into a series of steps. Think of it as a grid, where we plot points on the grid that match the waveform. The grid points are then encoded as data.

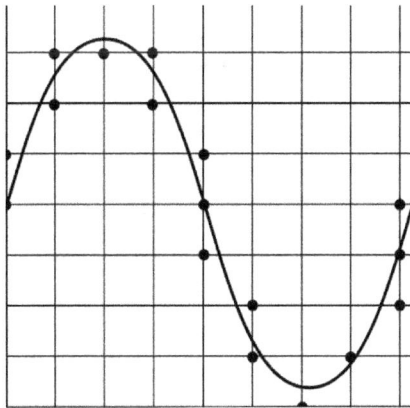

In this first example the grid is very low resolution, meaning there are few points of intersection on the vertical (dynamic range) and horizontal (frequency/time) axes. Once the system reconnects these dots during the digital–to–analog conversion for playback, it won't look like the original waveform. It's distorted, known as quantization error. But increase the resolution enough, such as we see in the next example, and there are sufficient data points to capture the nuances of what we can hear. *Sampling rate* is the number of "snapshots" taken per second of time (the horizontal axis) and *bit depth* represents amplitude (vertical axis).

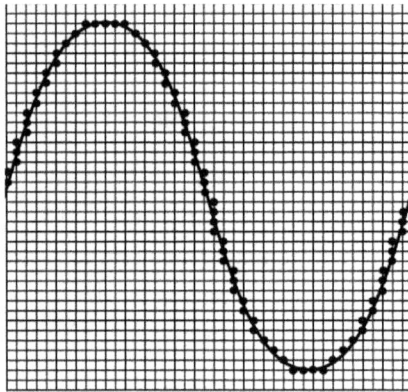

Harry Nyquest was a communications engineer in the first half of the 20th century who helped develop the roots of modern communications theory. Based on his work, the idea is that it requires at least twice as many samples per signal bandwidth to accurately capture it without generating unintended components. What happens is that if an audio signal gets sampled, or encoded, at less than 2x its highest frequency component, extra wave cycles will be introduced into the signal. This is known as aliasing, and so to prevent this the system will do two things: sample the audio at a rate at least twice the highest frequency while applying a low-pass filter to eliminate everything above the desired audio range. You've probably heard the number 44.1kHz associated with digital audio recording. This is the sampling rate specification for the compact disc, representing higher than twice the range of human hearing. The video industry adopted 48kHz for reasons video people understand (or did a long time ago), and improvements in technology have resulted in sample rates as high as 192kHz.

Go lower than 44.1, though, and it won't sound very good, especially for music.

Bit rate determines the resolution of dynamic range captured in each sample. Higher bit rates use longer digital words that can store more information. For example compare an 8-bit system (from the 80s) that has a word length of 8 digits. This gives you 256 unique values to encode a signal. It's quite crude and doesn't come close to matching the ever-changing analog waveform. A 16-bit system like the compact disc provides 65,536 values—much better. Current 24-bit processes, though, make the quantum leap to 16,777,216 values. Whereas for 16-bit recording we have to watch levels (don't overload, yet don't lose low-level signals), 24-bit recording gives you so much fine detail and range we can finally re-think old habits. Originating from the days of analog tape, the mindset was to push as close to the top as possible, avoiding distortion, in order to maximize available dynamic range without losing audio in the noise floor. That doesn't apply with 24-bit systems. Keep in mind that good levels are good levels, always, so make sure the meter is showing a healthy signal. For most everyday recording a 24-bit/44.1 or 48kHz setting will work fine. If storage is not an issue go for a 96kHz sample rate (twice as much data).

Most of the magic happens behind the scenes and you don't have to worry about it. A couple of issues, though, to be aware of. *Latency* refers to a delay between the original signal and what comes out of a digital system in real-time. This isn't much of an issue during mixing these days, unless you're running a ton of processor-intensive plugins on certain tracks. In a tracking session the problem is feeding a monitor mix during overdubbing. The computer is trying to play back tracks already recorded while inputting new audio to be digitized and routed back out to the headphones. Digitizing and processing audio takes time, even with current computer processing power, and the delay can be quite audible. Audio interfaces thus have an input/output monitor balance pot so you can dial it to whatever works for each situation. Some interfaces allow you to set up monitoring from the interface itself *before* it hits Pro Tools, keeping the computer out of the loop and therefore nearly eliminating latency issues.

Clock is the other term that comes up when working with multiple devices. The only way that long string of zeros and ones makes sense to the system is knowing exactly when each word starts. The problem is that this can get skewed, a phenomenon known as *jitter*. Timing is everything, and so a complete digital system much be referenced to a single, solid clock source. Set one device as the master; in some cases a dedicated clocking device must be added into the studio chain. Sounds crazy, but this actually affects the quality of the music. Some digital consoles, for example, sound better when clocked with an external, built-for-purpose clocking device.

There are several connection formats for digital audio. Home audio/theater components usually have an optical jack (Toslink) that transmits stereo and compressed multichannel audio (surround) using the S/PDIF format; some devices use the ubiquitous RCA jack instead. HDMI embeds multichannel audio with the video signal. ADAT Lightpipe (Alesis optical protocol), TDIF (Tascam digital interface format), and MADI (multichannel audio digital interface) are examples of multichannel connections. The groundbreaking trend at the time of this writing is converting digital audio to a format that can be transmitted across standard computer networks. Primarily gaining its footing in the live sound industry, the idea is to put a network jack (CAT5) on each audio device, connect it to a network switch, then use software to route audio data from any device to any other device on the network. This is revolutionary in that no longer do we have to worry about how to split a signal to go to two places. No longer does it take expensive equipment and cabling to connect signals to other devices, locations, etc. Need to set up a remote viewing room for an event that's sold out? No problem— just run a CAT5 cable from the network switch to the playback gear in the room and route from the computer. Recording studios with multiple tracking rooms and isolation booths required complicated cabling and patching systems; now just plug the mic into one of these interfaces and it will come up in the control room. Another example: at my college we often record musical events in the chapel on campus. This building is not connected to where our studios are located, so for decades the only way to do this was to have a complete recording system in the chapel. Now we can plug mics on stage into the campus network and see signals in our control room. My sound system at church runs off a single CAT5 cable from the console to the back room mic preamps and amplifiers. I just rewired our stage with CAT5 jacks in addition to regular XLR mic inputs. The possibilities are endless, and it'll be interesting to see where it all goes.

TWENTY
SOUNDS & SIGNALS

The heart of all the information we've discussed involves fundamental acoustics and electrical signals. Understanding these elements helps you better follow how we record sounds, route them through equipment, process them, and listen through speakers. Situations will arise in the studio and in live concerts where a solid understanding of acoustics and audio signals can help you determine possible causes and potential solutions. This chapter only attempts to get you started; there are many excellent books available that describe these concepts in greater detail.

Sound waves

How do sounds actually get from one point to another? If you see someone's mouth moving as they look at you, how does that covey anything meaningful such as speech? When we hear someone playing a musical instrument, how can we tell that it's a piano and not a snare drum? The answer lies in the vibration of air molecules—when an object vibrates, such as vocal cords, piano strings, or a saxophone reed, it changes the atmospheric pressure around it. These are relatively minor pressure changes—nothing like what happens when a hurricane comes through. But they are enough that when the vibration follows a particular pattern, it can convey information when received by our ears or by a microphone.

What's a pattern? Depends on the vibrating object. If it's a drumstick hitting a drum, then the sound pressure variations are fairly random and sound like noise to us. If it's a flute playing an A, then the sound waves follow a repetitive cycle—the particular waveform pattern unique to flutes repeats over and over until the playing stops. This is called a periodic wave, and all musical sounds feature this characteristic.

The sound itself is also distinguished by what the waveform looks like if you graph it out. Sound generated from a flute will set up a particular pattern of fluctuations different than that of a snare drum. These waveforms are built from simple sine waves—lots of them. You remember sine waves from geometry class, though you probably had no idea what they were used for. Now you have a practical application—sine waves, the most pure, simple form of vibration, combine in varying patterns to create complex sounds such as instruments, noises, speech, etc. When these patterns repeat over and over you get a musical sound; when they are random you get noise.

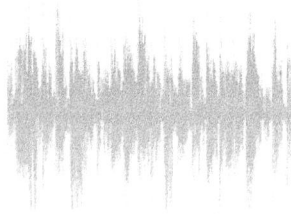

A complex musical waveform

In audio equipment such as consoles, recorders, and processors, there is no physical object vibrating (except for microphone diaphragms and speaker cones). When a microphone transduces an acoustic sound into an electrical signal, the pressure variations correspond to resulting changes in voltage. An increase in pressure equates to an increase in positive voltage, and a decrease in pressure is represented by negative voltage. Thus the electrical signal is analogous to the original acoustic sound (thus the term *analog* in audio recording). This can be seen by graphing a simple sine wave—the positive curve going up represents the original acoustic pressure increase, and the curve going down into negative territory comes from an acoustic pressure decrease. This constantly fluctuates around the zero point, which is equilibrium (no signal), just as the AC voltage in your house constantly fluctuates between positive and negative. With audio signals, however, the fluctuation is much more complex and interesting compared to the plain 60Hz repetitive cycle in your power lines.

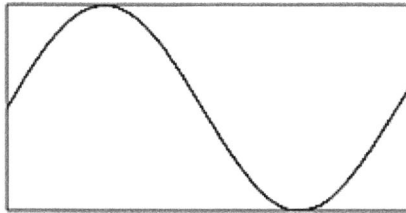

Simple sine wave where upward curve is positive voltage, down curve is negative

As you probably know, these vibrating patterns must oscillate (vibrate) between 20 and 20,000 times per second to be heard by humans. This is known as frequency—number of cycles per second, which roughly equates to our sense of pitch (high or low). Translating this to musical terms, particular pitches are based on specific frequencies, such as the standard tuning note of A = 440Hz. A musical octave can be heard every time the frequency of a sound doubles, meaning that if you play a 40Hz tone, then double that to 80Hz, it will sound like the same pitch, only an octave higher. Audio equalizers are built with this in mind, where graphic EQs feature filters set at octave intervals, third-of-an-octave intervals, etc.

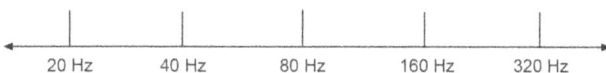

| 20 Hz | 40 Hz | 80 Hz | 160 Hz | 320 Hz |

Frequency doubles with each octave

Different frequencies have different wavelengths—low frequencies will have long waveforms (30 feet and longer) while higher frequencies possess much shorter waveforms (fractions of an inch). This is important when you consider issues of isolation in the studio, frequency response, and control of acoustics in a studio.

How much the variation moves from equilibrium (amount of pressure change) is referred to as amplitude. This is what gives us a sense of loudness, and in electrical signals running through your equipment it means how high or low your signal strength is. Moving a console fader up increases amplitude, moving it down decreases it. If a signal is too high in strength for a circuit it will distort; all audio equipment (including our ears) have a limited dynamic range, or capability of handling amplitude ranges. If a signal is too low, then you increase the amount of noise that is heard in the sound, so the engineer must balance signal levels pretty closely to get the best, cleanest sounding signal possible.

One more important concept for now—phase. Phase involves a time relationship for sound waves. Of particular interest in audio is when two identical, or nearly identical, soundwaves are shifted slightly in time from each other, then combined in a mixing console. Due to the normal algebraic summing of the wave components when you add two or more signals together, the varying positive and negative curves are now skewed somewhat and result in a different waveform. The difference depends on the amount of time shift which will affect different frequency components. The result? Your sound will change somewhat, usually resulting in a hollow, less-full sound. Why care about this? When you use more than one microphone at a time in the studio you run into phase issues. When you take a recorded track and process time delays and other effects you run into phase changes to your sound. When you set up for a live concert you must deal with phase relationships into your mics and in the hall due to wall/ceiling reflections. The most simple phase problem is when you wire your stereo speakers backwards on one side, or when a mic cable has been wired incorrectly. It's a big deal, it's difficult for novice engineers to comprehend, and it takes time to understand. Listen to your tracks as you record and see how it changes when you add that second mic to the mix—you'll eventually get the picture.

Characteristics of sound waves

Cycle

One complete sequence of a sound wave before it begins repeating itself. This refers to periodic waves, which are musical.

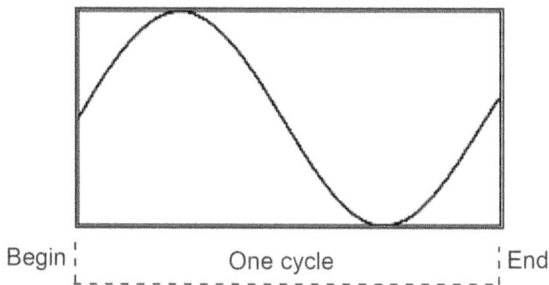

Begin   One cycle   End

Amplitude

Vertical distance of the wave from center line (zero or equilibrium). Relates to perception of loudness, although it's not exactly the same thing.

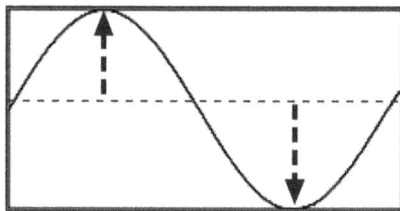

Frequency

Number of cycles completed per second (bandwidth). Relates to, but not the same as pitch. The audio bandwidth that humans can hear is between 20Hz–20kHz, diminishing on the high end as your hearing deteriorates over time. Keep in mind that this frequency range is only a small portion of the entire radio spectrum that includes radio, tv, microwave, etc.

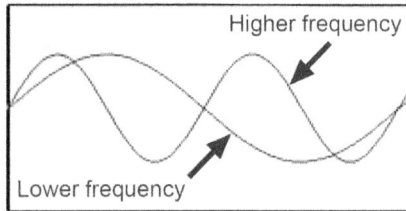

Wavelength

Physical length of the sound wave from one point in the cycle to the next identical point. Notice how the diagram looks identical to that of cycle. The difference is that cycle is measured in time and wavelength is a physical dimension. Low frequencies are longer while higher frequencies become progressively shorter.

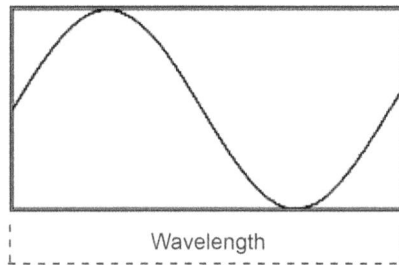

Phase

Measured in degrees ($360°$ = full cycle)

- *Phase cancellation*: $180°$ phase shift which results in total cancellation of both waves.
- *Phase shift*: time delay between two identical waves. Anything other than o or 180 will affect the sound by altering the harmonic structure.

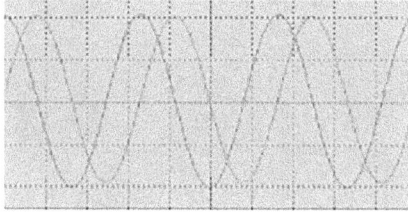

Harmonic content

All sound waves are made of a particular combination of sine wave components (harmonics).

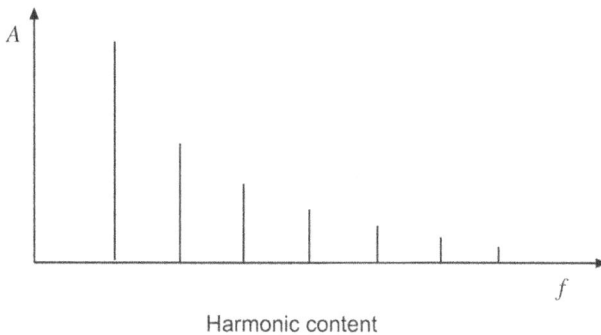

Harmonic content

Harmonics make sounds distinctive

Number and relative strengths of individual sine waves comprise a sound and determine tonality or timbre.

- *Fundamental*: lowest frequency in a complex wave. Determines basic pitch of the note.
- *Overtones / Partials*: higher frequencies in complex waves.
- *Harmonics*: overtones which are multiples of the fundamental.
- *Timbre*: perceived tone quality based on harmonic content. Identifies trumpet sounds from flutes, etc.
- *Noise*: contains all frequencies and has irregular waveform.

How exactly do sine waves add up to something more interesting than a test tone? Here is a single sine wave (test tone):

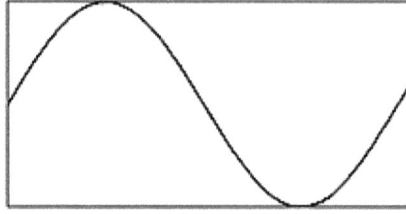

Next, we'll add a second sine wave at twice the frequency of the first. The graph below shows both individual waves.

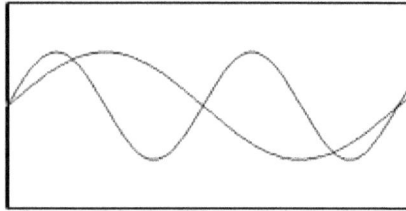

What happens when these two waves are combined? This results in a complex wave that does not look like a sine wave anymore. It also begins to have a more interesting sound, though it has a long way to go to sound like music and other noises that we're used to hearing. The darker wave in the graph represents the resulting complex wave.

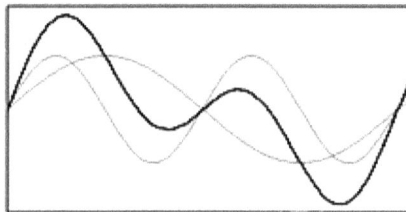

Finally, we've added a great many sine waves that, at certain frequencies and amplitudes, add up to a square wave (the dark line). This can be intentional as when using a synthesizer to reproduce a certain sound, or it can be a problem such as when you overdrive an amplifier or recording medium. When this happens, the original waveform has a higher signal level than the system can handle, so the extreme high and low amplitudes are chopped off (clipping). The result is a bunch of new sine wave components that weren't in the original sound and are now altering the sound in an unpleasant way (distortion).

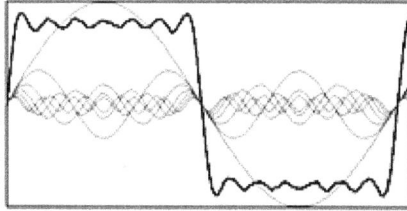

Here is an audio example of a waveform that was recorded too high, clipping the system. You can see the flattened top and bottom, which resembles the square wave described earlier.

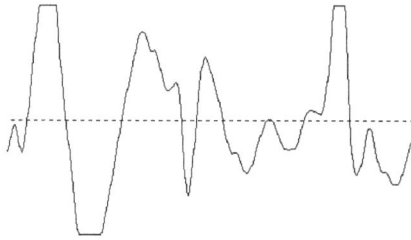

Why is this important?

Knowing frequency content of various instruments helps with setting EQ. It also helps you during mic selection—if a sound has mainly low frequencies (kick drum), then you need a mic that can adequately capture and reproduce that range. Understanding the issues of over-driving audio systems can help identify and prevent unwanted distortion.

Other basic acoustics concepts

Envelope

All sounds have a volume shape over time—how the sound rises and decays back to silence. We can alter this with dynamics processors (compressors and gates), and we control it when we use synthesizers to reproduce or create sounds. A percussion hit will have a very rapid attack and quick release, while a violin has a more gradual volume envelope. An organ has an instantaneous on, fairly steady sustain, then immediate release when you take your finger off the key.

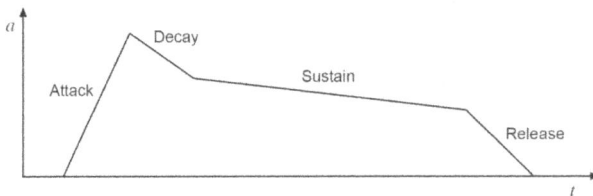

Velocity of sound in air

The speed of sound is different in various materials or mediums. We most often are concerned with how fast sound travels in air, though studio architectural design requires knowledge of how fast sound can be transmitted through various building materials. Live reinforcement engineers have to work around time delays between sound coming from the main speakers and directly from stage; they can calculate how long it will take for the original sound on stage to reach the audience, then program in specific delays into the loudspeaker system to compensate. Here's a comparison of sound velocity in different materials.

- Air: 1087 ft/sec at 32°F (344 m/s) dependent upon temperature (+/- 1 ft/sec per °F change)
- Water: 1433 m/s
- Glass: 3962 m/s
- Steel: 5029 m/s

Reflection, diffraction, and diffusion

When sound hits a surface three things can happen:

- Reflect away from the surface
- Lose energy through absorption by the surface
- Transmit (radiate) through the surface into the space on the other side

Which of these occurs depends on the nature of the room's surfaces, meaning what the material is and what may be attached to the surface. Compare the difference in sound between a bathroom, which usually has very hard surfaces such as tile, and a living room that is filled with soft objects such as furniture, plants, and carpet. The bathroom is much more lively and reverberant. Selection of materials and construction techniques are crucial for controlling the acoustics in a studio, concert hall, etc.

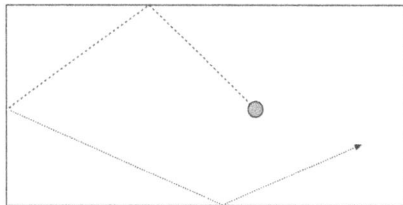

Hard surfaces reflect sound energy until eventually dying away

Sound reflections in the mid-high frequencies function similar to light waves in that the angle of reflection is equal to the angle of incidence. This is very helpful when working on room acoustics since we can't see what the sound waves are doing. Thus we can diagram where sound will go after it strikes any particular wall or ceiling in the room.

If sound reflects between two parallel surfaces, energy continues to bounce back and forth depending on the type of surface. Hard, heavy mass materials tend to reflect energy with minimal absorption, sustaining the ongoing reflections. Porous, lightweight materials will absorb some amount of the sound energy, converting it to residual heat.

This parallel back and forth is called *flutter echo* for higher frequencies and sounds like ringing; it's known as *standing waves* in the lower frequencies, which we'll get into a bit later.

Notice how the sound changes after you begin furnishing an empty room? Along with the objects added to the room, such as furniture, it's the changes in the wall and floor surfaces. When you hang pictures and tapestries on the walls and install bookshelves you create an irregular wall surface that scatters sound waves. This is a good thing, and we even employ specially-built units called diffusers to do this in a more efficient way in the studio. The idea is to distribute sound energy as evenly as possible throughout the frequency spectrum. It provides a more well-rounded ambiance, preserving a sense of liveness in your recording room that we need for music to sound natural.

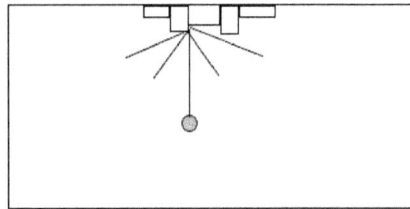

When sound energy hits an uneven surface, it scatters back into the room more evenly

Sound also bends around objects, more so in the low frequency range. This is how you can hear your neighbor's bass guitar down the hall, but not the high hat and cymbals. Sometime when you're bored and need some cheap entertainment, stand in front of a speaker, then move off to one side. The high frequencies will diminish, but the lows will stay solid. This can be a problem in the studio when trying to isolate instruments from each other during recording—bass frequencies will travel all over the room and leak into the microphones. It's also an issue when mounting and locating monitor speakers in the control room. Different placements in the room and around the equipment will give you very different sound responses. It's these invisible characteristics of sound that are so problematic for people trying to set up their own studios—

they do not understand what's really happening with the sound, and therefore their recordings suffer accordingly.

Human hearing

We won't get into a complicated discussion about how the brain deciphers sounds transduced by your ears. For now we'll focus on a few key aspects you should be aware of when working in the studio.

The main concept is that human hearing is non-linear. This means that you don't hear what you think you are hearing. In other words, there are various issues involved in how we perceive sounds. For instance, if you double the wattage of a power amplifier, you would assume that the volume would double. Not so. The actual perceived increase is only 3dB —barely noticeable. It takes 10 times as much amplification to make it sound twice as loud. Remember studying logarithms in school? This is where it comes into play, where changes in loudness (and pitch) is logarithmic and not linear. Later in this chapter we'll briefly introduce you to decibels and using logarithms to calculate various sound and signal levels—don't freak out, it'll be okay.

There are also differences in sound depending on how loud we listen. If you listen at lower volumes, you hear less bass and treble frequencies. If you crank it up very loud, you will hear much more bass and treble than usual. If you engineer your recordings based on these listening levels, they will sound quite different when played back in various situations. There is a graph that shows us how this works (Fletcher-Munson), and the suggested solution is to listen around 85dB SPL. Find a sound pressure level app for your phone and experiment with various volume levels.

One other issue to remember is that our ears can be "overloaded" with excessive volume. Not only will this cause permanent hearing loss over time, but there are immediate implications as well. Your ears will quickly tire and lose perspective the longer you work. If you quit for the day and come back later, you'll notice how different your recordings sound. Engineers should take breaks regularly during long sessions. If you listen to very loud levels for any extended period of time, you will also begin to hear distortion at some point, but this is actually happening in your own hearing system. The concept is exactly the same as what happens when you overload a power amplifier (or any audio component). The signal hits maximum dynamic range and any level over that point gets clipped. This essentially creates a square wave, adding a host of unrelated sine components. The resulting waveform is different and adds noise that is usually unwanted and unpleasant. Our ears do the same thing. Early in my career I was working with a producer who liked to monitor at really high volumes. It wasn't long before I was hearing distortion in the monitors, but it wasn't from any signal levels I had running through the console. It took a moment to realize it was all in my head, so to speak, but in a very real way. Knowing I had set everything up okay before this kicked in allowed me to complete the session without worrying about what was being recorded, but it was a weird experience.

Perception of direction

The chapter on stereo microphone technique discussed briefly how humans hear direction of sound. Using differences in amplitude and time of arrival between left and right, we can tell where sound sources are located. This is also used to create a sense of direction when mixing in the studio. Pan potentiometers and digital delays follow these two principles when placing images in various stereophonic locations. Different stereo miking techniques follow these principles in capturing a sense of stereo space from the stage source.

Perception of space

Humans get a sense of the room they are listening in through such cues as direct sounds and reflections. This was described in the chapter on time-based processors. Once a sound is generated, the listener will hear the direct sound straight from the source, then begin to detect early and late reflections from nearby surfaces. These are not distinct echoes, but very closely-spaced reflections in the millisecond range. The longer it takes for the brain to detect an initial reflection after hearing the original direct sound, the larger the room is perceived to be. We can therefore use time-based processors to create any type or size room for a particular recording.

Audio signals (electrical)

What is an audio signal?

We can take an acoustic sound wave and convert (transduce) it into an analogous electrical signal. Microphones take care of this function, and then we can route these signals through mixing consoles, signal processors, and recorders. Eventually these signals will be converted again into acoustic waves so we can hear them through speakers. The point is that the basic principles we discussed for acoustic waves apply the same to signals—it's the same audio information, only conveyed in a different transmission medium. Keep in mind we're talking about analog signals, not digital data converted for your DAW; that data contains the same information, but operates a bit differently.

With this in mind, you now know that audio signals possess characteristics of frequency, amplitude, phase, etc. These are very important concepts to understand when working with audio. Again, read a good text that explains in-depth the nuts and bolts of audio signal theory. For now, we'll provide just a few examples of why you should pay attention to this stuff.

Wonder why old recordings seem to lack that luster and clarity of newer recordings? Older recording equipment lacked the frequency response of modern technology, meaning higher and lower frequencies on either end of the audio spectrum simply could not be captured and reproduced. Most obvious is the standard "telephone" sound, which we now mimic as a vocal effect by rolling off the highs and lows, leaving only the mid-range. Knowing the frequency ranges of various instruments and voices helps you find EQ settings that fit each particular part. We select different microphones for certain miking situations based on their individual frequency response and sound—they're intentionally made to sound unique. So, you'd most likely put a large-diaphragm mic on a kick drum in order to match that instrument's low-end frequency range.

Signal amplitude refers to how much juice is flowing through your circuits. Over-drive your microphone preamp and it'll distort. Crank up too much EQ and you can overload the EQ circuits. Record too hot on your recorder and it'll distort. If you set signal levels too low, however, you get a lot more noise added to your sound. Connect a piece of gear into another that's not matched correctly and you'll have problems getting the levels to work. Notice some equipment that has a level change switch on the back, say between +4 and -10 or -20? That's there for a good reason, and you need to know which level your overall studio is running so you can purchase and connect equipment that will work correctly.

We've mentioned phase issues that exist acoustically when you use multiple mics in a room, and the results show up when mixing those signals on the console. The most common situation is miking a drum set. When you're recording a drum set and have several mics running at once, start listening to one mic at a time, gradually building the drum mix. At some point you might notice that the kick, snare, or something else thins out. You just added a mic that's in adverse phase relationship with the mic on that particular instrument. Either move one of the mics a little (might not take much), or try flipping the phase button on one of the mic channels on the console. Use your ears or install a phase/polarity meter for a visual indicator.

We can run into phase problems electronically as well. Take a track, duplicate it while adding a short delay, then pan together and listen for how the sound thins somewhat (depending on delay time). Changing the delay time will dramatically change the sound. You can also run into electronic phase problems with incorrectly wired cables. For example, mic cables have three wires inside, two for signal and one for ground. The two signal wires must be connected to the correct pins, and if these are reversed at one end of the cable you've got a complete 180 degree polarity shift. That's a problem, so you should check your cables with a cable tester to be sure.

Logarithms

In the world of audio we constantly need to measure acoustic sound and signals running through circuits. We need to know how loud a concert is, how many speaker cabinets and amps will be needed to run the show, how to set our meters and recorders in the studio, and find out the before and after affects of an increase or decrease in sound or signal level.

So, how do we go about measuring signal levels? When you look at a console or DAW you'll see meters labeled in units called dB. So when the vocal peaks around 2dB, what exactly does that mean? Nothing by itself. Decibels are meaningless unless they are referenced to something that actually does mean something. Unlike a yard, meter, or mile, which have very specific quantifiable values, decibels have no value in and of themselves. What we do is compare them to values that are based on specific measurements.

In acoustics, we use a dB SPL meter to measure how loud a concert is. The result, say around 110dB SPL, indicates that the concert acoustic level is 110 decibels louder than the threshold of hearing (0dB), which is the lowest point humans can detect sounds. In electronics, we measure a vocal signal running through a channel by comparing the current voltage (or wattage) to an established standard reference level for audio equipment. So, relative to what has been established as 0dB for voltage readings, the vocal signal is occasionally higher by 2dB. Decibels are essentially expressing a ratio between two known quantities, providing useful information that can be employed in a variety of situations.

Why use dB? Decibels are logarithmic, which is useful to us in a couple of ways. First, the range of acoustic intensity is tremendous—lots of really big numbers that engineers don't want to use in daily life. The same is true for voltage readings and such. It's much easier to find a system that reduces this to a more manageable way of working with changes in SPL or signal levels. So, even though the difference in sound pressure between the threshold of hearing and a space shuttle launching in your driveway is almost beyond comprehension with lots of digits and decimal places, we can easily say that the difference is around 150dB (give or take). Just remember that every 10dB increase is about equal to a doubling of perceived volume, so small numeric increases in dB actually represent significant changes in intensity or level as you go up the scale (that's the log part of it). Every time you double your sound system amplifier wattage you're not getting a sound that's twice as loud, but rather a small increase in perceived volume. That's why it takes several stacks of amps to run a large sound reinforcement system.

The other reason to use logs is that we hear logarithmically. Remember that every time frequency is doubled it sounds an octave higher, no matter where you are on the spectrum? The actual quantity of frequencies between an octave of 20Hz–40Hz and an octave of 10,000Hz and 20,000Hz is very different. The bandwidth is 20Hz in the lower octave and 10,000Hz in the higher octave. However, each octave sounds like the same interval, whether it's in the high or low range.

20 Hz	40 Hz	80 Hz	160 Hz	320 Hz

Frequency doubles with each octave

Logs provide a system of measurement that allows the use of reasonable mathematic figures and comparisons that match our non-linear perception of hearing. If you recall from math class in high school (yeah, right), a log of a value (X) is the number which, when applied to 10 as an exponent, produces that value (X). For example, the log of 100 = 2, because two 10s, 10x10, equals 100. These days we just use the formula and log functions on our phones. Here is one such formula for calculating dB increase in amplifier output, where we are comparing the original power level ($P_{reference}$) to the new power level ($P_{measured}$).

$$dB = 10 \log P_{meas} / P_{ref}$$

In case you're interested, the decibel system comes from the original telephone pioneer Alexander Graham Bell. Telephone systems actually used the unit *bel*, but that's too large for audio, so we measure in increments of one-tenth of a bel, or decibel.

Different types of dB measurements

Decibels are difficult enough for novices (and many professionals) to comprehend, but it gets even more complex due to the different decibel applications and references. Essentially decibels are used for the following applications:

- Measuring sound pressure level (or intensity) to see how loud it is (acoustic measurement).
- Measuring signal level (electrical or digital measurement).
- Measuring changes in level (acoustic or electronic).

You should understand that when we use a dB indicator, it is based on some other physical measurement such as voltage, wattage, or acoustic pressure. The equipment measures the physical quantity, then converts that into dB indicators for ease of use. Decibels always provide a comparison between two physical quantities, whether it's an established reference level or the difference between two current measurements. Here is a brief outline of the more common applications.

Measurement applications

dB SPL

Sound Pressure Level (acoustic). We measure pressure change of acoustic waves in comparison to the threshold of hearing (zero). We can also measure sound intensity, which uses a different physical measurement. Higher SPL levels correspond to higher perceived volume.

dBu (dBv)

Pro audio equipment voltage. This is the most common measurement for analog audio signals, and the 0dB reference is equal to 0.775 volts. In your equipment, however, 0dB on the meter actually equates to +4dBu, or 1.23 volts, if you measure with a dB or voltmeter. Professional equipment runs higher "behind the scenes" for better audio quality, thus the reason why 0dB VU is actually +4dBu.

dBm

Pro audio equipment power. You'll see this specification in many equipment manuals, but it's not used as often. Originally based on telephone system equipment that ran on 600ohm loads, dBm is used for power measurements (wattage) and not voltage. It's only good when measuring in 600ohm loads, however, and you don't use this for calculating power amplifier output levels.

dB FS

Typically referred to as 0dBFS (full scale), this represents maximum level in a digital system. If you look at meters on analog recorders and mixers you'll see 0dB just past the halfway point with readings beyond this, in the red, up to +3dB. 0dB is optimum, but not maximum. On a digital system, however, 0dB is as high as it goes before it runs out of bits to encode the signal, resulting in immediate, harsh distortion that will ruin your day.

dB LUFS

We discussed this in the mastering chapter; LUFS is a loudness measurement based on perception and psychoacoustic analysis. Similar to dBFS, it's a digital scale where zero is maximum. All music streaming services, as well as the broadcast industry, use loudness units for establishing levels, the idea being a more consistent volume experience for listeners.

Useful tips to keep in mind

- Doubling of power (wattage) only results in a 3dB increase—

not much in perceived volume. So if you just got your parents to replace your 50W guitar amp with a cool 100W amp, you will be quite bummed.

- Doubling of voltage results in a 6dB increase.
- Doubling of distance results in 6dB decrease.

This last principle comes in mighty handy in the studio when miking multiple instruments in the room. Move the sources farther apart along with their mics; if a mic is moved twice as far away from another source, the leakage into that mic will drop a bit less than 6 dB or so. You'll get a cleaner sound overall.

How do decibels provide measurements?

By comparing a measured, quantifiable value against a reference value.

$$\text{n dB SPL above the threshold of hearing}$$
$$\text{dBSPL} = 20\log \text{SPL}_{meas} / \text{SPL}_{ref}$$

$$\text{n dBu above reference of 0.316 volts}$$
$$\text{dBu} = 20\log V_{meas} / V_{ref}$$

$$\text{n dBm above reference of 0.001 watts}$$
$$\text{dBm} = 10\log P_{meas} / P_{ref} \ (P = \text{power or wattage})$$

Here's an example of how the math works (you'll have to experiment with your phone app or calculator to make sure you know how to use the log functions correctly).

If a shuttle launch produces a sound intensity of 20 watts/m², use the standard formula and plug in the reference value of 10^{-12} watts/m² as follows:

$$P = 20 \text{ watts}/m^2 \text{ (the shuttle)}$$
$$PR = 10^{-12} \text{ watts}/m^2 \text{ (threshold of hearing)}$$
$$dB = 10 \log P/PR$$
$$dB = 10 \log (20^1/10^{-12})$$
$$dB = 10 \log 2^{(1+12)}$$
$$dB = 10 \log 2^{13}$$
$$dB = 133$$

Another example:

$$\text{Measured sound pressure level} = .075 \text{ dynes}/cm^2$$
$$\text{Reference level} = .0002 \text{ dynes}/cm^2$$
$$dB \text{ SPL} = 20 \log .075/.0002$$
$$dB \text{ SPL} = 51.5$$

Here's how to figure the dB increase with your new guitar amplifier:

$$\text{Original guitar amp was 50 watts}$$
$$\text{New guitar amp is 100 watts}$$
$$dB = 10 \log 100/50$$
$$dB \text{ increase} = 3.01$$

Note that with wattage (power) measurements the formula uses 10 log, whereas with voltage and sound pressure measurements 20 log is used. This is true for both acoustic and electrical calculations.

Tools to measure SPL and signal level

Meters on the console, DAW, or recorder

We've mentioned audio meters from time to time. There are two primary types, and they provide a visual indication of signal level. However, they operate quite differently and will affect your recording decisions accordingly.

VU (Volume Unit). These have the little needle that swings over toward the right. VUs provide an averaged value that corresponds to our perception of volume. It will not show all the peaks and dips, the same as the fact that our ears don't register very brief bursts of signal (transients). You can overload something without realizing it, so consoles typically also provide peak LED indicators that will flash when you're pushing close to the danger zone.

Peak Reading. These are usually step graph indicators or simple LEDs. They are designed to respond to any instantaneous signal change, and will therefore show all level peaks. Sometimes they will also feature a "peak hold" function, meaning it will keep the highest indicator lit for a little longer so you don't miss any transients along the way. For DAWs, make sure you know what mode your meters are running; it makes a difference.

And remember that volume level is sourced from signal level, but you cannot know how your levels are running through the system merely by how loud it is in the room. Always watch your meters to make sure your recording levels are good, that you're not overloading the mix bus, and so on.

Sound pressure level meter

SPL meters measure acoustic sound pressure levels as described earlier in this chapter. They actually measure the changes in pressure using dynes/cm2, then convert this to a decibel reading that makes sense to engineers. They're pretty easy to use, but you have to understand the weighting network system to use them properly.

If you read up on Fletcher-Munson curves, you'll know that this graph shows us that we hear lows, mids, and highs differently at different volumes. To get an accurate reading on an SPL meter, you must adjust the internal weighting network for the general volume level you will be running in your situation. These are referred to as A, B, and C weights with the following applicable SPL levels:

- A: Referenced to 40 Phons (levels of 40dB or lower)
- B: Referenced to 70 Phons (levels in the 70dB range)

- C: Referenced to 100 Phons (levels of 100dB or higher)

Other audio meters

There are several other devices used to measure audio for various applications. Standard multimeters can be used to directly measure voltage, current, or resistance in an audio circuit. Specialized audio tools can generate test signals and measure dB comparisons, frequency response, noise levels, impedance, etc. These can range from $10 for a cheap multimeter to several thousand dollars for high quality maintenance equipment. Much of this can also be accomplished with phone apps for very reasonable prices.

TWENTY-ONE

STUDIO DESIGN

No matter how much expense and effort you put into buying good gear and learning how to use it, you'll keep tearing your hair out because your mixes don't sound very good. Some of this is because you're not quite there yet as an engineer, but one of the biggest factors in getting a good recording is the room itself. As soon as sound is produced from an instrument or vocalist, it begins radiating into the surrounding space. How this process develops over time before the sound dissipates determines whether you capture a clean or muddy sound, distant or close sound, or something that doesn't sound at all like what you hear standing there in the room.

Everybody has heard the difference between a bathroom, living room, gym, and large auditorium, but few understand what's happening. The ideal recording studio distributes sound evenly throughout the space, maintaining a smooth frequency response that doesn't emphasize or lose certain frequency ranges. Music in the room should possess clarity while having enough ambiance to give it some life. But within these guidelines are tremendously variable preferences, depending on the types of music and activity being recorded. An orchestra requires a large, reverberant space for all the acoustic instruments to blend into what we recognize as an orchestra sound. A rock band still needs ambiance and liveness to the sound, but much less reverberation and tighter control in order to preserve clarity for each part. At the far end of the spectrum is spoken word, for which a very controlled, reverberant-free environment is necessary for clarity of speech. Studio facilities are designed with their primary purposes in mind, and many facilities have the ability to provide different acoustical characteristics for each type of session.

I was once hired to finish the installation of a new recording studio. When I walked into the tracking room, the door closed behind me like an airlock on the Starship Enterprise. The room was completely dead, no ambiance at all, and it sounded boomy. The problem was too much high-frequency absorption—carpet on the floor, a solid line of sound panels around the walls, and really thick acoustic ceiling tiles. And this was from a professional audio systems design company, who had no excuse for such a terrible room. But most DIY engineers have little idea how to treat and fix the rooms they have, much less design and build an adequate room from scratch. Let's hit the high points of what the issues are and how you can improve your lot in life. Be warned, though. This is a very intricate, complex field of science. While there are some basics that will help in most situations, specific understandings and solutions require the assistance of an acoustical engineer who's well experienced in this type of work.

Fundamental room acoustics

Once something generates sound in a room, those sound waves propagate (travel) around the room in various ways. The size, shape, and surface treatments of the room determine how this happens, and thus how it sounds.

As we briefly discussed in the sounds and signals chapter, propagated sound progresses through three stages:

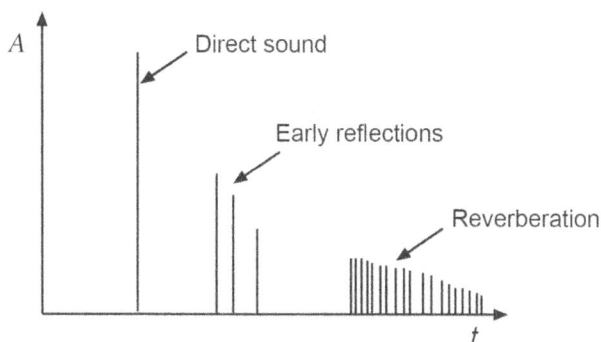

Propagation of sound in a room (simplified)

Direct sound is what you get straight from the source, with no reflections from nearby surfaces affecting it. The closer the mic is to a sound source, the more direct sound it will capture.

Early reflections are the first waves to be reflected from nearby surfaces such as walls, ceiling, or floor. Unless it's a really big area, they will not be heard as distinct echoes, so you can't sit there and say "Oh, I just experienced an early reflection. That was pretty wild". Your brain, however, is able to detect these various waves and decodes that into a sense of how big the room is. This comes from the delay time between the original direct sound and these early reflections, so this is primarily how you can tell whether you're in an auditorium or small club. A longer initial delay indicates a larger room, because the reflective surface is farther away from the listener and takes a bit longer to arrive. If the delay between reflections is long enough, we hear distinct echoes. This occurs in very large, reflective rooms such as gyms as well as outdoor stadiums.

The major problem with surface reflections is that they can destructively affect your sound. When a vocal travels directly into the mic while at the same time bouncing off the nearest wall, it combines in a way that affects intelligibility and clarity (recall the phase cancellation issue we've discussed a few times). You've got two nearly identical sounds arriving at the mic at different times (the reflection takes a bit longer due to its longer path length), then combining to create a waveform quite different from what it's supposed to be. The difference in arrival time determines which frequencies are primarily affected, since frequencies have different physical wavelengths. Experiment with this by taking a track in your DAW, duplicate it, then nudge one of them over just a tad. Keep sliding it back and forth and you'll hear the resulting timbre change.

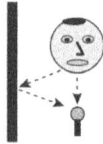

The third stage after direct and early reflections is reverberation. Sound reflections continue to travel throughout the room and bounce in various directions. If left untamed, they rapidly multiply and develop into a dense sound we call reverberation. If the walls, ceiling, and floor are mostly hard, flat surfaces, you get lots of reflected sound and more reverb. Larger rooms have the potential for more reverb because there's more airspace for sound to propagate without quick attenuation.

The goal for all of this is balance. You need direct sound combined with some early reflections and a bit of reverberation for most music recording. The issue is controlling each of these in a way best suited for the type of recording being performed.

Absorbing sound energy

Imagine you're a stuntman on a movie set and they ask you to jump off a building into the street below. What would you rather dive into, the pavement or an inflatable mat? The asphalt is no good because it doesn't give, so all that energy gets pushed back into you (most likely breaking bones, though I've never tried it myself). The inflatable mat, on the other hand, is soft and cushiony, so it collapses and absorbs energy from your fall. Sound bouncing around a room is no different. A gym built from concrete block with steel structural beams overhead will be highly reflective since these materials are very rigid and absorb very little sound energy. When you move into a new house or apartment you'll notice that the empty rooms sound hollow and echoey. Once you add furniture, plants, pictures on the walls, window drapes, bookshelves, carpet, and cats, it sounds much more subdued. There are two main principles here, one of them being that these items absorb sound energy. You need soft and fuzzy materials to absorb sound; they're not as dense as brick, concrete, and metal, and so instead of reflecting back into the room, soundwaves get trapped in the material where the energy dissipates. The other principle is that these items diffuse sound around the room. We'll get to that in a second.

High and mid frequencies

You've probably seen churches, classrooms, and auditoriums that have sound panels mounted on the walls. Most of the time these are compressed fiberglass panels covered with fabric. This material is an excellent, low-cost sound absorber and has been used for years. The disadvantage is that they're mostly limited to high and mid frequencies; low frequencies have much longer wavelengths (they're bigger) and require a different approach. Rooms that only have these types of panels on the walls will typically sound somewhat subdued and boomy; the high frequencies that provide ambiance and life to your sound are gone while the low frequencies roam untouched around the room. This requires a balanced approach in terms of materials and installation methods.

Fiberglas panels are available in 1-inch, 2-inch, and 3-inch thicknesses. 1-inch panels are really thin, so they're only effective for high frequencies. Thicker panels easily handle high frequencies, but are also able to absorb a bit farther down in the mid-frequency ranges. Try to use at least 2-inch panels on your walls; 3-inch is even better for lower frequencies. You'll have lots of choices of fabric colors to match your room design, and you can even order panels with artwork or photos screened on the fabric. If you're handy with basic carpentry, you can make these yourself. The fiberglass most often used is Owens Corning 703, and you want a fabric that's fireproof rated, such as Guilford of Maine. Build a frame to hold the panel and to wrap the fabric around.

Fiberglas panels should be installed on bare wall areas. Try to cover a good amount of the wall surface for it to make a difference (at least 50%); they can be spaced out from each other a bit or clustered in blocks depending on the kind of look you prefer. The key is to not leave large blank spots. Installation is a breeze; you can use construction glue and impaler clips. These metal clips are flat with sharp points sticking out. Screw the flat base of the clip to the wall, then slide the panel onto the impalers. Don't cover all your wall space, though, as that will deaden the room too much. We'll balance this with diffusers, which we'll touch on soon.

Low frequencies

Low frequencies are a different animal because they have really long wavelengths and much more energy than high frequency sounds. These waves therefore don't dissipate as quickly as higher frequencies; excessive low frequency energy in a room will cause it to sound boomy, so special absorption techniques are required. The other quirk with low frequencies is that because they retain so much energy over time, if a wave bounces off a wall back toward the opposing wall, it will actually combine at points, sustaining energy in that band and causing what is known as *standing waves*. Now, standing waves, also known as *modes*, aren't as evil as many people have made them out to be. All rooms have modes; the issue is when a close grouping of modes occurs due to the dimensions of the room.

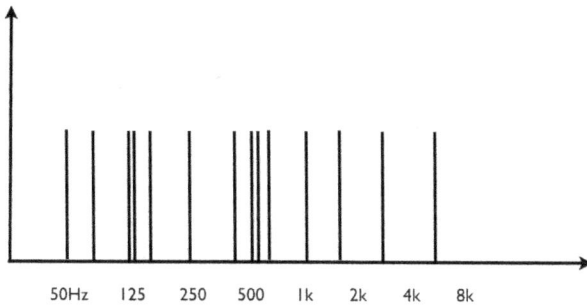

Room modes clustered around 125 and 500Hz

Remember that different frequencies have different wavelengths. When the room dimensions coincide with specific frequencies, the back and forth reflection between walls will line up and reinforce sound energy. Studio designers take great care in selecting room dimensions in order to reduce this issue. The worst case scenario is a room where all three dimensions are equal: length, width, and height. If a room is ten feet in each direction, you'll have a reinforcement of the same low frequency energy in all dimensions, creating a tight cluster of identical modes. The goal is for all dimensions to not be equal or multiples of each other. You'll still have a specific frequency for each measurement that can potentially cause an issue, but at least they won't double or triple up.

So how can you determine what the modes are for a particular room? Several websites provide mode calculators, or you can do it yourself with a standard formula: F = v/2d.

F is the modal frequency we're looking for, *v* is the speed of sound, and *d* represents each measured room dimension. We can look at it like this: F = 1130/2(dimension). Here are a couple examples, the first being a terrible choice with all dimensions related. Frequencies in bold indicate modes that are closely related among at least two dimensions.

Room length is 24ft, height is 16ft, and width is 8ft.

	1st harm	2nd	3rd	4th	5th	6th
$F_L = v/2d = 1130/2(24)$	24	48	**72**	96	120	**144**
$F_L = v/2d = 1130/2(16)$	35	**70**	105	**140**	175	**210**
$F_L = v/2d = 1130/2(8)$	**71**	**142**	**213**	284	355	426

This one is better; room dimensions are not quite multiples. Note how the bolded frequencies don't line up as closely; the modes are spaced out a bit more, which helps. This room isn't perfect, but better than the first.

Room length is 15ft, height is 11ft, width is 8ft.

	1st harm	2nd	3rd	4th	5th	6th
$F_L = v/2d = 1130/2(15)$	38	76	114	152	190	228
$F_L = v/2d = 1130/2(11)$	51	102	153	204	255	306
$F_L = v/2d = 1130/2(8)$	71	142	213	284	355	426

Smaller rooms suffer from standing waves higher up in the audible range, so larger rooms are a bit easier to work with. But, once you calculate the modes you can purchase or build traps tuned to each specific frequency. This takes a bit more research and thought than merely putting up acoustic panels on the wall, so either take some time to figure it out or hire an acoustician who can perform acoustic measurements of the space and make recommendations.

To absorb low frequencies we mount a semi-rigid membrane (plywood or something similarly stiff) in a frame with airspace behind the membrane. When low frequencies hit the membrane, it gives a bit, absorbing some energy from the wave. Have you ever caught a baseball line-drive with your bare hand? If you hold your hand still to catch it, it stings. A lot. But if you pull your hand back just as the ball gets there, it doesn't hurt so much. You've absorbed much of the energy, like a shock absorber. That's how a bass trap works. The airspace between it and the wall behind acts like a sponge, soaking up more low frequency energy.

The disadvantage of bass traps is that they're much thicker and heavier than standard acoustic panels. Unlike the acoustic panels, though, which need to be placed according to where sound is reflecting in the room, bass traps can go nearly anywhere since low frequencies are largely non-directional. You're already familiar with this if you have a sub-woofer in your home theater system; you can put it anywhere in the room and for the most part you can't tell where it is during the movie.

Good bass traps can be fairly expensive, and even though most smaller rooms don't need many, it can add up quickly. If your room is drywall construction, you've already got some low-end absorption built-in as this material flexes—just like a trap. Building your own is a viable option. Remember, the basic concept is to have a rigid membrane, such as plywood, framed and mounted several inches off the wall. Build your frame on the wall, screw the plywood onto the front, then seal the sides of the frame. Don't add cross-pieces behind the plywood—it must only be attached along the edges so it can flex. For more broadband absorption attach thick panels of Owens Corning 703 (compressed fiberglass boards) to the wall inside.

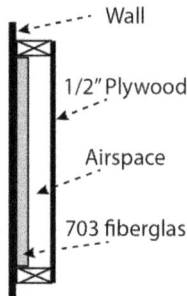

Wall
1/2" Plywood
Airspace
703 fiberglas

Traps can be built for a specific frequency as discussed above, or they can handle a broadband range of frequencies simply by angling the membrane from the wall. An easy way to do this is to build the trap in a corner. This creates a variable-depth absorber, which works very well. Corner traps are also advantageous because they're out of the way.

Some low frequency absorption can be achieved simply by mounting standard acoustic panels so that there is an airspace of a few inches between the panel and the wall. Panel manufacturers usually offer offset impaler clips just for this purpose, or you can build simple frames that hold the panel away from the wall.

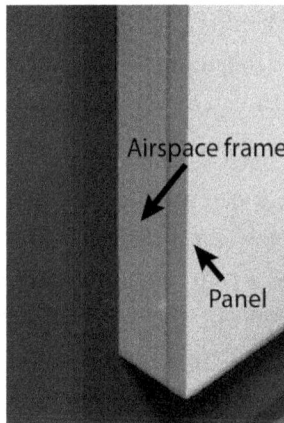

Distributing sound evenly

Absorbers aim to capture sound energy and prevent it from reflecting back into the room. But, if we absorb all our sound we're robbing ourselves of desirable ambiance that makes music come to life. Since we also do not want sound to reflect uncontrollably around the room, we use materials and objects that control these reflections without killing them entirely. This is called *diffusion*, meaning that when a sound hits an irregular surface, the various frequency components in that sound are scattered back into the room in different directions. The trick is to do this evenly and gradually. Hard reflections merely bounce straight back into the room and cause problems. Diffusion tames these reflections in a way that's acoustically pleasing to the ear.

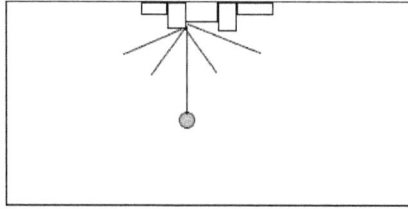

We diffuse sound by breaking up the flat surface, creating various angles and depths that affect different frequencies. How does this work? Since frequencies vary in wavelength, when these components hit various objects and angles they'll bounce back in various directions dependent upon how each component relates to the size and angle of the obstacle. Sounds complicated, and much of this is based on some pretty high-level math. But, you don't need to worry about any of that, because we can buy products that take care of this for us. You could also build basic diffusers yourself once you see some examples; just keep in mind that the math does indeed matter for maximum efficiency.

Diffusers come in a variety of sizes and designs; go check out acoustic manufacturer and dealer websites to get some ideas. They're thicker than standard absorption panels because they must have varying surface dimensions to do their trick. Some models work well on your walls interspersed with regular absorption panels; some combine both absorption and diffusion in the same unit. Yet others are designed to drop into standard ceiling grids, which is easy and convenient.

Speaking of ceilings, most office buildings with ceiling grids have standard "acoustic tiles". These are better than drywall, but hardly ideal. I would replace most, but not necessarily all, of these with diffusers like in the photo. This prevents direct reflections back down into the microphones, proving a nice diffusive ambiance in the room.

So for your recording space, look at any flat, bare walls and consider a balance of diffusers and absorbers. One option is to put absorbers on one wall with diffusers on the opposing wall. Or you could combine absorbers and diffusers on each wall. It's okay to have some blank space —you don't have to cover every square inch—but avoid large patches where there's nothing to break up or absorb sound energy. Many manufacturers of acoustic materials have very helpful diagrams, articles, and online calculators to point you in the right direction and even sell complete kits for various size rooms. Even if you don't buy one, you can easily see the concept of what they're trying to achieve with each setup.

A quick word about floors. Most engineers hate carpet and love wood. Wood is terrific, which seems contradictory considering it's a hard reflector. You wouldn't want an entire room of wood, but some of the best studios have at least one wood wall and wood floors for a terrific ambiance. I didn't have the budget for a hardwood floor for our studios at the college, so we built a large drum platform with layers of 3/4" plywood (so it won't flex), filled it with sand underneath, and we left the top plywood layer unfinished for a really nice, raw wood reflection. It sounds terrific for drums, acoustic guitars, and most anything else I put on it.

Summary for DIY treatment

- Absorb highs and mids with acoustical panels on the wall.
- Balance these with diffusion, also mounted on walls.
- Install diffusers in the ceiling grid (or attach directly if it's a solid surface).
- Install bass traps in the corners.

Control room design

In the early days of recording studios, the practice was to build large control rooms. Small rooms sounded bad, and nobody had quite figured out what to do about it yet. Early reflections were the primary issue along with significant low frequency modes high up in the audible range. The large monitor speakers typically used in those years would fill the room with sound, causing reflections from all surfaces. Engineers gradually learned how to shape the room and add absorption, but it was a while before practical solutions came along that made smaller rooms usable.

Two main concepts were crucial for this, the first being the near-field monitor. These smaller speakers were designed to sit on or very near the console bridge, focused directly at the engineer with little room interaction. Along with this came further development in controlling reflections, primarily through improved acoustic materials and the development of a reflection free zone. The idea here is to ensure that no direct surface reflections enter the engineer's listening position. By angling, or splaying, the front-side walls outward, any sound energy that strikes these surfaces is angled away from the engineer. This follows the basic physics principle where a wave reflects from a surface at the same angle it hits. As sound energy is focused along the sides toward the back of the room, careful balance of absorption and diffusion ensures sufficient ambiance for listening purposes while preventing direct reflections back toward the engineer.

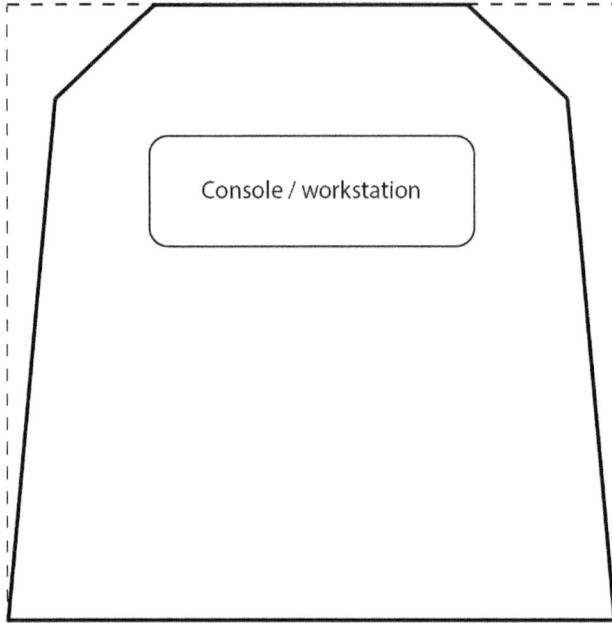

Along with acoustic benefits, splayed walls also provide for clever ways to divide up a large space into separate areas such as the control room, tracking room, isolation booths, and even bass traps. But, while they're a common feature in professional studio design, it's not an absolute necessity for setting up a good studio. Most available rooms that are turned into recording spaces are rectangular. This is a recipe for flutter and standing wave issues, but these are easily corrected with carefully placed absorption and trapping.

In the diagram below, absorption panels are installed directly behind the console to reduce reflections from behind the monitors. Panels are also placed to either side of the mix position, reducing side reflections. Bass traps are installed in the corners, while diffusers are on the back wall, providing ambiance while preventing direct reflections back into the mix position.

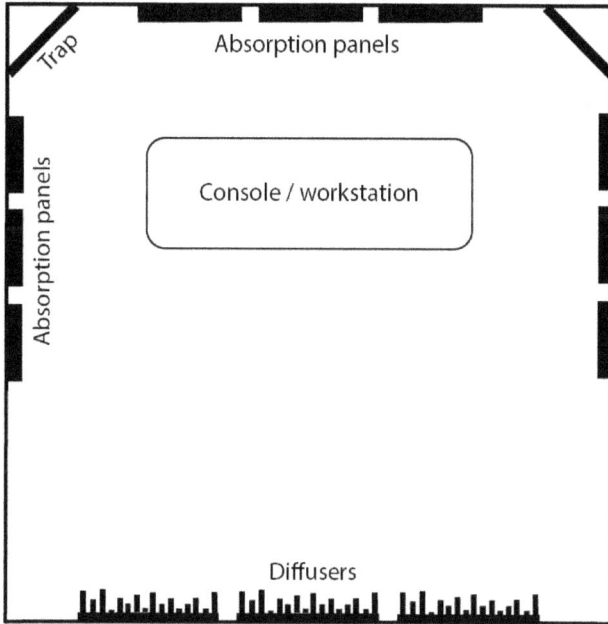

The overall goals for control room design are:

- Sonically neutral: must accurately represent what's on the recording
- No direct reflections in the mix position
- Symmetry of room layout (to preserve stereo imaging)
- Control of rear reflections
- All of these require application of absorption, diffusion, and geometry

Monitoring issues in the control room

When I was in college, several of us students kept complaining about how bad the monitor speakers sounded. Not that we were any kind of expert, but they just didn't sound right. Eventually we figured out it wasn't the speaker—it was where they were placed. Few young engineers give a second thought as to where to put their monitor speakers. Most simply place their near-fields directly on the console meter bridge, but sound radiating from the speaker components goes directly down to the console surface where it's reflected back up toward the engineer. This is the same issue as miking a singer close to a nearby wall or window, where the reflected sound causes phase interference with the direct sound. Not running a console? Putting monitors on the same table with your laptop does exactly the same thing. The secret is to locate the monitors just *behind* the console bridge, so that downward radiating sound waves strike the very top of the bridge, or even behind it, and therefore reflect away from the engineer. For DAW users, buy good monitor stands and set them behind your table. Once you can visualize reflection angles and see how angle of reflection equals angle of incidence, you can rough in a good spot. A quick acoustic measurement with pink noise can provide an indication of how much console splash you've got, meaning the readout will show lots of dips and notches if it's a big problem. Just move things around until you get it as flat as possible (it'll never be perfect).

The other issue is reducing reflections from nearby walls, such as behind the console and speakers. Careful acoustic treatment as discussed earlier can take care of this; just make sure there isn't a bare wall right behind (or close beside) the speakers.

Along with positioning the speakers behind the console meter bridge in our college studio, we also discovered the equilateral triangle rule. The distance between the two monitors, as well as the distance from each monitor to the engineer's mix position, must be exactly the same. This ensures even coverage from the monitors as well as accurate stereo imaging.

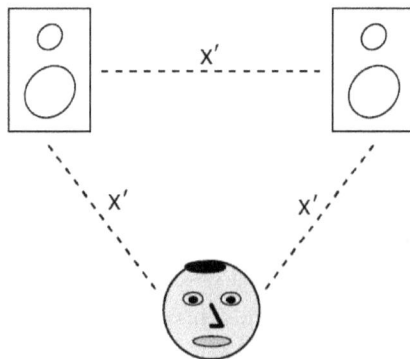

Finally, it's important to select the best quality monitors you can afford. Don't ever use home consumer speakers as they don't offer accurate frequency response. Small monitors suffer from inadequate bass response. Adding a sub can help, but this requires careful calibration to ensure a smooth crossover between the sub and mains. Headphones are problematic because they enhance effects and stereo imaging, and in general sound different than speakers in a room. Having said that, many consumers listen to music on earbuds or headphones, so comparing your mix on different devices is important. Good monitors are expensive, but probably one of the best investments you can make in your studio.

Tracking rooms

Tracking rooms are not as complex as control rooms; the primary goal is to create an ambient environment, free of uncontrolled reflections, that matches the type of music or voice work predominant for that facility. Walls should not be covered with too much absorption, but rather broken up with a combination of absorption and diffusion. Wood is a highly favored surface material for both floors and walls as it provides a very rich, warm, musical ambiance. Balance is key, and you may have to experiment a bit over time to get a room sound you're happy with.

Many studios provide variable acoustics, where one end of a room might be more live than the other. This might be the difference in carpet vs tile, or perhaps more absorption on certain walls. Adjustable acoustics can be designed into the room with drapes that can be opened or closed. "Window shutters" hinged on the wall is another clever method; one side is hard and reflective, the back side is absorptive. Swing the shutters one way or the other to change up the nature of the space. Very easy to do and quite effective. A roll of carpet can be quickly spread out on the floor, covering reflective tile. Gobos with opposing reflective, absorptive, or even diffusive surfaces can be positioned accordingly.

With all the attention to acoustics, don't forget that musicians need to see each other. Many studios have multiple rooms for tracking, so the vocalist is stuck in an isolation booth, the rhythm section is in the main room, and the drummer gets shoved into a corner. Large windows or sliding glass doors are necessary, even though they are highly reflective and create reflection issues. You can design around this, so be sympathetic, especially if you've never experienced what it's like to be on the other side of the glass.

Isolation vs absorption

Most people confuse isolation, which is preventing sound from transmitting *between* two spaces, with absorption. Absorption only deals with acoustics within a space. Keeping those trucks, birds, and obnoxious teenage drummers out is a completely different thing, so installing fiberglass panels on the wall won't help at all. It can also be very expensive, although there are reasonable techniques that can improve isolation between rooms in your facility.

The keys to isolation are mass and decoupling. Denser surface material with a higher mass factor will flex less, thereby not transmitting sound waves. Preventing two materials from touching effectively decouples the transmission. So layering wall materials with airspace in between each side can be an effective method for improving transmission loss (TL). Walls in your home or office typically consist of a single layer of drywall on either side of wood or steel studs. Drywall flexes easily, thereby transmitting sound waves directly to the other side, but the airspace in between helps slow this down. To increase mass, you can add a second layer of drywall directly to the existing wall on both sides, but it's even more effective if you first nail furring strips (1x3 boards) flat on the wall, running vertically 16" off-center, filling in between with insulation. Attach a layer or two of drywall to this; the increase in mass and the insulation-filled airspace provides perhaps 10dB TL improvement. By the way, if the studio is on a building's second floor or higher, the same trick can be done to the room's floor, increasing mass and airspace to reduce transmission to and from the floor below. If it's a ground floor facility there's no reason for this.

For new construction, build a double wall consisting of two completely independent wall structures parallel to each other. This includes two sets of studs with at least two layers of drywall on the outer sides. Fiberglas insulation is installed in between, and the airspace between both walls is preserved by completely decoupling everything. This technique can provide around 55 or 60dB of TL, which is not insignificant. How much do you need? A typical studio environment should run around 20 —25dB (A weighted). Use an SPL meter and measure the ambient noise inside the room, then subtract. This difference is what you need to build into the room.

Once upon a time there was a major recording studio with two tracking rooms adjacent to each other. The first time they scheduled two concurrent sessions it was chaos...because the designer forgot to cut the concrete floor between the two rooms. And this is the reason why you might not be able to fix all the isolation issues at your facility. It's one thing to add drywall and such, but you've got a floor that runs the length of the building. The ceiling structure is tied to building joists and beams that don't care how your rooms are divided underneath. All of this transmits vibrations quite easily. And forget "acoustic tiles" in the grid; these are only designed for absorption (and not particularly great at that). These issues can be dealt with, but it's expensive, invasive, and most studios cannot afford it. You may have heard of floating floors and rooms; the idea is to build each room's floor independently on top of isolators, essentially shock absorbers or rubber gaskets, that absorb vibrations. The edges of the floor run near, but do not touch, the exterior walls. Interior walls are built on top of rubber gaskets and connect to the main building frame with isolators. Ceilings are hung with devices that are essentially heavy springs. We won't spend any more time with this here, but at least you understand a little about the issues involved.

Doors, cabling, and other necessities

Let's briefly mention a few other odds and ends that you'll need to deal with. Doors are terrible because it's a hole in the wall; the main reason you hear folks clowning around in the hallway outside your room is because of the crack under the door and perhaps around the edges.

Anytime you see light around a door or window, sound is flowing unobstructed. Hollow doors such as you might find in your home or office are insufficient for stopping sound; the door flexes and merely transmits lows and mids through to the other side. Install a solid door, put an acoustic seal on the bottom (lots of good options to buy; don't use a cheap rubber gasket from Home Depot), and line the edges of the door frame with rubber weatherstripping so it seals tightly when shut.

Glass flexes quite easily, and so windows are a terrible isolator. The general practice is to mount two panes of glass with an airspace in between. Ideally the panes are different thicknesses, and if mounted at an angle from each other it further reduces direct transmission through the space. That's why some studios have hugely thick walls and windows between their control rooms and tracking spaces. If you're building the window frames yourself, line the frame with rubber for the glass to rest on, then seal the edges. If it's an exterior window and you don't need the sunlight, build over it so it becomes part of the surrounding wall. Think twice about this though...sunlight is a precious commodity when you're stuck in a studio for hours and days at a time.

Vents from the building's HVAC system are difficult to deal with for two reasons. It practically begs for sound leakage between rooms because ductwork is an open channel that connects each room. The other issue is air noise when the system is running. The larger the duct and vent, the quieter the airflow, although you may not be able to do anything about this. Ducts can also be lined with materials to absorb sound. The cheapest solution? Turn the air conditioning off when recording.

Cabling, including electric, must connect all the rooms. This requires punching holes in the walls, so make them as small as practicable and seal with spray insulation. Offsetting the hole on each side of the wall avoids a direct shot; sound enters the hole from one room, then runs into a solid surface.

In the control room there's a ton of cabling that runs around connecting the console, outboard gear, recorders, and so on. These rooms are often built with a raised floor, allowing space underneath to run cable troughs. The idea is to make it easy to pull cable at any time without tearing your room apart. And you'll have to change it over time, so make sure the troughs are large and easy to access.

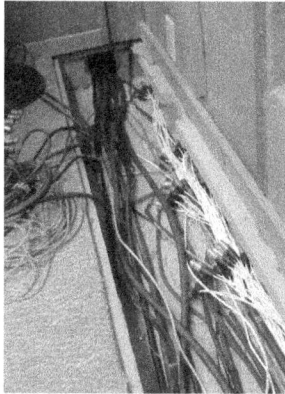

In my early years of college teaching I built a new recording studio in our building, but did not have a budget for acoustic treatment. The room was bare and sounded awful, so one day in acoustics class I brought in a couple rolls of house insulation, all the spare cardboard boxes I could find, and some staple guns. I instructed one team of students to staple the insulation along the top edge of the room while another couple of teams stapled the boxes to the walls in random configurations. The bottoms of the boxes were facing away from the wall. It looked terrible, but it was a rudimentary (desperate?) solution that provided high frequency absorption (the insulation), low frequency absorption (the boxes). The random placement of various size boxes, stapled to the wall at different angles, provided diffusion to scatter sound waves. Did it sound better? You bet. Did it sound great? Not a chance.

The moral of this story is that at least *something* can improve your room if it follows basic acoustic principles. I now have really nice studio facilities, but it took time and intermediate steps along the way. Start with something and then keep working at it.

TWENTY-TWO

RECORDING FOR MUSIC EDUCATORS

As an educator I have a particular interest in encouraging opportunities for students to explore and be creative. We don't do enough of that in school (or life), and music teachers are in a great position for facilitating this. I have the privilege of working with music educators in our master's program at the college, and we get to talk recording, go in the studio and try a few things, then send them off to make a difference. This chapter is designed with teachers in mind, reviewing applications and the how-to that takes what we've been discussing throughout the book and focuses it toward the music classroom.

Why do recording in school?

There are several reasons to incorporate audio recording in a school context. Music instructors often record their concerts. Students need audition recordings to send off for contests, scholarships, and so on. Recordings of rehearsals and private lessons provide valuable instructional feedback and also maintain a record of progress for assessment purposes. Students get to record their own songs and perhaps sell them as fundraisers. And who doesn't need to set up a simple PA system so the audience can hear the soloist or narrator? Let's step through each of these, remembering all the stuff we've covered that provides the foundation.

Recording ensembles

All you need is a simple recorder and a pair of decent microphones. A stereo microphone, which has two capsules embedded in the unit, is the easiest approach as you don't have to worry with positioning two individual mics correctly. Stereo mics will have a special adaptor cable that divides into two connectors for plugging into the recorder. A small hand-held digital recorder works fine, such as the models from Zoom, Tascam, or Yamaha. These usually come with built-in mics, which are pretty good on the better models. For external mics, plug them into the recorder's microphone preamplifiers (XLR jacks), turn on phantom power, and set a good recording level. Make sure the flash card has plenty of storage available; 24-bit recording at 44.1kHz requires 15MB per stereo minute, so a one hour concert is pushing 1GB.

When you're done, connect the recorder via USB to your computer to transfer the files. In your software, pan this stereo file left and right, insert a low-cut filter set around 75Hz (unless it's an orchestra or other source with low frequency instruments), and insert an EQ. Try attenuating the low-mids to reduce mud and boost slightly in the upper-mids for presence. Add a compressor set for 2:1 or 3:1 with just a couple dB of gain reduction, but don't overdo it. Finally, experiment with a peak limiter to help maximize the overall loudness, but again don't crunch it too hard. See the mastering chapter for more on this.

It's the mics that make the difference, so spend as much as you can for good ones. They must be condenser models for ensemble recording. If you buy two mics, as opposed to a stereo mic, make sure they're identical —don't mix and match. See the chapter on stereo microphone techniques for mic placement, but the idea is to locate the stereo pair several feet in front of the ensemble and several feet higher than the conductor. Go high enough so the mics can point forward and down into the group rather than facing directly into the front row. Always listen to the balance you're getting and adjust accordingly. If it sounds too echoey or roomy move the mics closer; if it sounds too close and in-your-face move them up or back. If you're using the built-in mics on the recorder the challenge is positioning the unit in the right spot while setting input levels, starting and stopping the recording, and perhaps feeding power to it. They're designed to mount on a standard video tripod, so if the stand is tall enough this will work fine. Otherwise buy an adaptor that will mount a video device on a microphone stand. Height is the issue, needing to reach at least 8–10 feet in the air, preferably higher. Start the recorder, raise the stand into place and have the group play a loud section of the piece, then check the levels to make sure it's not clipping.

Audition recordings

These typically involve a single student, perhaps with a piano accompanist. If it's only the one player, put up a single condenser microphone a few feet away in front of them. Listen to how the instrument sounds and move the mic around for a natural response without too much room tone. If it's a soloist with piano accompaniment, use a stereo mic in front of them to capture a stereo image. Listen for balance between each player and shift the musicians and/or mic accordingly (you can't change this later). We're not multitracking in this situation, just capturing a natural stereo image of the performance. Also consider the room they're playing in; stay away from flat, blank walls and windows. Sound panels on the wall are your friend, but bookshelves or instrument cubbies can help break up direct reflections. A large, empty band room can be a terrible environment unless it's acoustically treated, so if the recording sounds washed out with too much reverb try to find an office or classroom. The closer the mic, the less room sound, so listen for a balance you like. By the way, for simple recordings a USB mic is very inexpensive and will work just fine. Plug it directly to your computer, open the audio software, and select this mic as the input source.

Recording student bands

Many of your students write songs and would love the opportunity to record them. You don't need a sophisticated, expensive recording studio, though. Sure, it's nice, but remember the basics for getting a great recording: decent musician, decent instrument, good mic placement. You could put an empty can of green beans in front of a great musician and it'll sound awesome. So the first step is to help students learn to play as well as possible, know how to set up and tune their instrument, and learn the songs.

Now, let's assume you already have a stereo mic or pair of condensers for recording concerts and rehearsals. If this is all you have, simply use this to capture the entire band. Arrange people in the room so as to balance everybody toward the mics. If you have additional mics, put the stereo mic over, and perhaps slightly in front of, the drumset, positioned to capture a balance of the entire kit. A large diaphragm dynamic in the kick drum hole will fill out the bottom end, with smaller dynamics on snare and toms. Inexpensive drum mic kits are available that will serve the purpose quite well. Any other mics in the same room will get lots of drum sound, so consider using direct boxes whenever possible, spreading players out, or overdubbing parts. Run bass guitars through a direct box and don't use a bass amp. Put guitar amps in a closet or different room with a long cable so the player can stay in the room with the others. Acoustic guitars will need overdubbing later unless you use the pickup with a DI (not the best sound). The microphone technique chapter has miking suggestions for these as well as other instruments.

You'll have to decide whether the group can handle overdubbing, which is good practice for them, or needs to play together at the same time. Listen to each mic signal so you know what it's picking up; make adjustments now as you can't easily fix it later. Remember that most mics are directional, so point the backs of each mic toward other instruments in the room to reduce leakage. Spreading players away from each other also reduces leakage. Finally, think about room acoustics—try to tame the environment with sound panels, blankets, instrument cubbies, winter coats draped on music stands, and so on.

All of this could be run through a simple mixer and recorded to the same stereo recorder used for concerts. But it's much better to record individual tracks. This requires an audio interface with enough mic inputs to handle the number of parts you want to record at once. Shop with caution, though. Product manufacturers will claim x-number of inputs, but most of these are not mic inputs. It must be a preamplifier with an XLR jack. DAW software on the computer can record individual tracks, then allow you to work with each part later to fine tune the mix. Garage-band works great and is free for Mac users; Apple's Logic Pro is not unreasonably priced for education users if you want to get more serious about all this. Many interfaces come bundled with DAW software, so this is not an obstacle for getting started.

For overdubbing, just plug headphones into the interface if it's close enough or buy a headphone splitter. Connect the headphone jack on the interface into the splitter, then run individual headphone lines to each musician. The easiest method is to send them the same monitor mix you're getting from the DAW mix bus, but if the interface has more than two outputs you could set up a separate cue mix. See the tracking chapter for details.

Final file formats

Once your editing, mixing, and mastering is finished, bounce or export the files as both a WAV and mp3 file (AAC if you're uploading to iTunes). This gives you a high quality master that can be used later as needed along with the small, portable mp3 for sharing. If you're making CDs use the WAV master. Again, review the mastering chapter for more on this.

PA systems

It's often the teacher struggling to get the programs printed, fix broken instruments at the last minute, and then have to set up the PA system the day of the big show. I have another book that covers running sound for events, but the short version is that you need a couple main speakers on stands, a mixer, and cables to connect everything. My recommenda-tion is to get active, or powered speakers. These have built-in amplifiers, eliminating the need for separate amps. Get a small digital mixer with enough mic inputs. The ideal setup, though more expensive, is a complete system featuring a digital mixer that connects to a mic input unit on stage via a single ethernet cable; the speakers are then connected to the stage unit. Digital mixers allow you to store settings, plug in an mp3 player, or even lock out curious folks who have no business turning dials. Analog mixers can work fine for this, though, so don't despair if that's what's available.

Plug in a mic and have someone talk or sing. Powered speakers will have level controls on the back, so turn these up perhaps halfway or so for now. Turn up the main mix fader on the right side to where it says zero or unity (this isn't the very top). Turn on the mic channel and push the fader up, then slowly turn up the mic preamp level control at the top of the channel. Make sure no one holds a mic close to the front of the main speakers or you'll get feedback. The more mic channels that are on and turned up the greater chance for feedback, so turn off anything not used for the moment. EQ can help with this, but if you're unfamiliar with this process just pull the fader down a bit. Place the speakers out to the side and in front of where your mics will be located. Use dynamic mics only as condensers are so sensitive they'll pick up everything around them and increase feedback issues. The only exception to this would be a podium mic, which is usually a condenser. Turn on the low-cut filter on all channels, set to around 80Hz, to reduce rumble and stage noise. Finally, a generic EQ setting with some low-mid attenuation (to reduce mud) and a slight upper-mid boost (for clarity) can work well on most vocals.

But my school has no money!

This is an all-too-common problem. The good news is that there are lots of inexpensive and free products that will get you started. Be patient and chip away one piece at a time. Perhaps you have an active parent booster program that can help, or local grants to apply for. Ask the local music shops if they have anything B-stock or used they'd like to donate. Be careful, though, about assuming you can sell recordings to make money. If they only feature student compositions you're good, but any commercial songs are off limits unless you're willing to pay the license fees (see www.harryfox.com for details).

My advice is to prove your ideas worthy of attention—don't wait and whine, but get started and do something, anything, to get the ball rolling. If it's well done, people will start to notice. Students will get excited and join up, attracting the attention of your administration, and suddenly you're a hero. Or something like that.

ABOUT THE AUTHOR

Dr. Barry R. Hill is a professor of audio engineering, instructional designer, and writer. He is Director of the Audio & Music Production degree program at Lebanon Valley College in Pennsylvania and a member of the National Academy of Recording Arts and Sciences (Grammys), Audio Engineering Society, and the Themed Entertainment Association. For fun he produces *The Themed Attraction Podcast* with a few friends who work in the themed design industry.

Dr. Hill holds degrees in Instructional Design from The Pennsylvania State University, Music Technology & Interactive Media from New York University, and Music with Recording Arts from the University of North Carolina Asheville. He can be contacted at www.barryrhill.com.

www.ingramcontent.com/pod-product-compliance
Lightning Source LLC
Chambersburg PA
CBHW071320210326
41597CB00015B/1288